ULTIMATE CAR SPOTTER'S GUIDE 1946-1969

Tad Burness

Published by

 **krause
publications**

700 E. State Street • Iola, WI 54990-0001
Telephone: 715/445-2214

Please call or write for our free catalog.
Our toll-free number to place an order or obtain a free catalog is 800-258-0929
or please use our regular business telephone 715-445-2214
for editorial comment and further information.

Library of Congress Catalog Number: 98-84622
ISBN: 0-87341-629-5

Printed in the United States of America

Table of Contents

Acknowledgments

Though most of this book is derived from pictures and information in my own reference files, special thanks are certainly due to the following individuals, for their kind help in contributing hard-to-find pictures, ads, brochures, spec sheets, etc. (Car makes in which certain individuals specialize are in parenthesis after their names.)

Bill Adams (Mopar)
Ronald Adams
Jim Allen
Jeff Anderson
Warren J. Baier
Larry Blodget (Ford)
Jim Bollman
Edwill H. Brown (Hudson)
Emmett P. Burke
Ken Buttolph
John A. Conde (AMC)
Virginia Daugert (Kaiser-Frazer)
Howard DeSart (Mopar)
Jim Edwards
Jim Evans (Ford)
Fred K. Fox (Studebaker)
Norm Frey (Mopar)
Jeff Gibson (Edsel)
Bruce Gilbert
Dick Grove (Studebaker)
Allan Gutcher

Albert R. Hedges
Larry C. Holian
Alden Jewell
Elliott Kahn
Mike Lamm
Daisy Lowenstein
Bob Martin
Keith Marvin
Carl Mendoza (Hudson)
Dave Newell (Corvair)
Walter F. Robinson
Charles Rowe
Jay Sherwin (Willys Jeep)
Mark Simon
Phil Skinner
Bob Snyder
Bryon Stappler
Ty Thompson
R.A. Wawrzyniak
Robert Zimmerman

Thanks are due, also, to the following corporations and associations, for helpful details and cooperation:

American Motors Corp. (now part of Chrysler Corp.)
Autophyle Magazine
Buick Club of America
Cadillac-LaSalle Club
Chrysler Corp. (Mopar)
Contemporary Historical Vehicle Assn. (CHVA)
Corvair Society of America (CORSA)
Crosley Automobile Club
DeSoto Club of America
Edsel Owners' Club
Fabulous 50s Ford Club of America
Ford Motor Co.
General Motors Corp. (& Divisions)
Hudson-Essex-Terraplane Club (HET)
International Edsel Club

Kaiser-Frazer Owner's Club
Milestone Car Society
Oldsmobile Club of America
Pontiac-Oakland Club
Society of Automotive Historians
Studebaker Driver's Club
Vintage Corvette Club of America
WPC/Chrysler Product Restorer's Club

An Enthusiastic Introduction

This new book is designed to be a helpful tool for you; with it, you can identify any American production car built between 1946 and 1969! ... and it's perfect for settling arguments with friends about the various characteristics of these cars.

For many years (since the 1970s), my various other *American Car Spotter's Guides* have been used by car buffs, collectors and others as a handy reference source -- because of the many illustrations. Because the demand for these books has far exceeded the supply, I've received countless requests to re-issue the old editions.

However, since I did those earlier books in the 1970s and 1980s, I've acquired so much more in the way of good pictures and information that it calls for a completely new, revised edition: Specifically, this new *Ultimate American Car Spotter's Guide*.

By adding many more pages to this new edition, and removing a few pages of pre-1946 cars and of recreational sport-utility vehicles (which could be used in other books), we're able to give you so much more that you never saw in earlier editions -- making use of plenty of recently-found material!

Much gratitude is due to certain readers of my books and syndicated "Auto Album" newspaper series, who have from time to time sent me helpful pictures and information I could use in books or in "Auto Album." Without their kind help, this book would not have been possible.

In this new book you'll find more and better interior and instrument panel views, as well as much more data on prices, horsepower, production figures, etc. Though this information was painstakingly researched for months, from many sources, such sources can sometimes disagree (which makes automotive historians want to tear out their hair!). Original prices, too, may differ -- depending on where the car was sold or whether it was sold early or late in the model year. Some West Coast prices have been included, in addition to the Midwest and eastern prices most reference sources use.

Browse through and see your favorite old Fords, Chevys and Plymouths, those unforgettable bullet-nosed Studebakers, the step-down Hudsons, bathtub Nashes, buck-toothed '50 Buicks, the Packards, Lincoln Continentals, Chrysler 300s, and all your favorite chrome-plated dreamboats and your old family cars from 1946 to '69!

Enjoy this book! I hope it's one you'll want to keep and look through again and again.

With best wishes,

Tad Burness
P.O. Box 247, Pacific Grove, CA 93950

AMERICAN MOTORS

Ambassador

$2404.~2968.
PRICE RANGE

232 CID
6
(155 HP)
287 CID
V8
(198 HP)
OR 327 CID V8
(250 OR 270 HP)

116" WB $2574.

990   CVT.

990   4-DR.

new DPL.
H/T →
$2756.

66

(68,084 BLT.)

INTRO. 10-7-65
(880 SERIES
ALSO AVAIL.)

HAS CHECKER PATTERN UPHOLSTERY

990

(FOR PRE-'66 AMBASSADORS, SEE **RAMBLER** .)

From $2515 to $3143*
(50,391 BLT.)

Six or V8

AMBASSADOR WAGONS

$3083.

The Red Carpet Ride.   new 118" WB

990 4 DR.
$2776.
$3/28.
(WEST COAST)

67

880
2-DR.
$2519.

RESTYLED

232 CID 6 (155 HP)
new 290 CID V8 (200 HP)
new 343 CID V8
(235 OR 280 HP)

DPL.

DPL

HOOD ORNAMENT
ADDED

DASH

DPL
$2958.

1967

DPL GRILLE
BEARS 2
RALLY LIGHTS

AMBASSADOR DPL CONVERTIBLE AND HARDTOP

(DPL CVT. IS
ONLY AVAIL. '67)

THE **NOW** CARS FROM THE
1967 AMERICAN MOTORS

AMERICAN MOTORS AMBASSADOR, DPL, OR SST

MODELS; NO MORE 880 OR 990 SERIES)

SEDAN *$3065.

WEST COAST PRICE

DASH

new DOOR HANDLES

$2820.

new SIDE SAFETY LIGHTS

new GRILLE

$2947.

DPL (INTRO. 9-26-67) **68** AIR COND. NOW STD. EQUIP.

NO MORE HOOD ORNAMENT UNTIL 1974

SST (ABOVE) has RALLY LIGHTS ON GRILLE.

$2820. ~ 3207. PRICE RANGE

(1969 MODELS INTRO. 10-1-68)

AMBASSADOR SEDAN, DPL and SST SEDANS, WAGONS and H/Ts AVAILABLE
$2914. ~ 3998. PRICE RANGE

(NO MORE AMBASSADOR MODELS AFTER 1974.)

new HORIZONTAL SHAPE FOR FRONT SIDE SAFETY LIGHTS.

REAR FENDER DETAIL

"Ambassador" NAME ON new GRILLE; HORIZONTAL ROW OF HDLTS.

$3,165 (SALE)

69

AMBER LTS. NOW IN NARROWER SLOTS

SEE ALSO **RAMBLER**

REAR

American Motors   **Classic**   *(1961 TO 1966)*

("RAMBLER CLASSIC")

(CONT'D. FROM "RAMBLER" SECTION)

intermediate size car   *112" WB*

WAGONS FROM $2888.

AVAILABLE IN "550," "770" OR "REBEL" SUB-MODELS.

2 SIXES and 3 V8s AVAILABLE. STANDARD ENGS.: 287 CID V8 (198 HP) OR 232 CID 6 (145 HP) 6.95/7.35 × 14 TIRES

770
CVT. $3065.

66

(THE FINAL "CLASSIC" LINE)

new GRILLE

(WHITE VINYL TOP ALSO AVAIL., 3-66)

new "CRISP-LINE" ROOF (Rebel/H/T)

$2972.

(MARLIN ALSO IN CLASSIC LINE, '65-'66.)

(SEE AMC MARLIN)

Rebel

(INTRO. 10-7-65)

(AVAIL. ONLY AS 2-DR. H/T)

(CLASSIC REPLACED BY THE **AMERICAN MOTORS** 1967 **Rebel**)

FOR 1965 and EARLIER "CLASSIC" SERIES, PLEASE SEE "RAMBLER" SECTION.

AMERICAN MOTORS
Javelin

(1968 TO 1974)

(INTRO. 9-26-67)

68

(56,462 BLT.)

**brand new,
8 cylinder, 280 horsepower**

343 C/D (290 C/D V8 ALSO)

(232 C/D 6 ALSO)

109" WB (THROUGH '70)

6.95/ 7.35 x 14 TIRES (THROUGH '69)

$2743.
(WEST COAST)
SST = $2848.

BASE PRICE = $2482. 2587. (SST)

DASH HAS DEEPLY-RECESSED ROUND GAUGES and CONTROLS

AMERICAN MOTORS *JAVELIN* $2512. $2633. (SST)

(INTRO. 10-1-68)

69

SCRIPT "JAVELIN" NAME NOW ABOVE GRILLE.

↗ SPECIAL

Big Bad Javelin

IN "BIG BAD ORANGE," "BIG BAD BLUE" OR "BIG BAD GREEN," WITH PAINTED FRONT and REAR BUMPERS, OTHER OPTIONS.

new "BULL'S EYE"

(40,675 BLT.)

(ASSOCIATED WITH JAVELIN SERIES THROUGH '74)

AMX A-8 SERIAL # PREFIX

(INTRO. 2-24-68)

A-9 PREFIX FOR 1969, MORE STRIPE COMBINATIONS.

$3485. ('68)

$3571. ('69) (WEST COAST)

ENGINES:
290 CID V8 (225 HP) OR OPTIONAL
343 CID V8 (280 HP)
390 CID V8 (315 HP)

EA. AMX CAR'S PRODUCTION NUMBER IS SET IN DASH.

97" WB

E70×14" TIRES

1969 MODEL INTRO. 10-1-68.

DETAILS OF HOOD

68-69

(new)

BASE PRICES =
$3245. ('68)
3297. ('69)

READILY IDENTIFIED BY UNIQUE DUAL WEDGES ON HOOD, EACH BEARING 5 PARALLEL LOUVRES.

6725 BLT. 1968
8293 " 1969

11

American Motors MARLIN

Marlin BY RAMBLER — Newest of the Sensible Spectaculars
(ANNOUNCED 2-65)

(1965 - 1967)

65

116" WB
232 CID 6 (155 HP)
OR
287 CID V8 (198 HP)
OR
327 CID V8
270 HP @ 4700 RPM

POWER DISC BRAKES STANDARD

10,327 '65 MARLINS BLT.

WIRE WHEEL DETAIL

EASILY IDENTIFIED BY UNIQUE FASTBACK "KNIFE-EDGE" REAR STYLING

7.35 OR 7.75 × 14 TIRES

$3143. f.o.b. and up

Marlin GRILLE
and SEATS (RECLINING)

Introducing excitement!
The swinging new man-size sports-fastback – MARLIN!

6557 2~DR. FASTBACK H/T IS ONLY MODEL AVAIL.

SEE ALSO:
RAMBLER

INTERIOR, THROUGH LONG SIDE WINDOW AREA

AMERICAN MOTORS MARLIN

(1965 - 1967)

6 CYL. OR V8

IN 1965, 1966, A PART OF THE CLASSIC LINE

112" WB (THROUGH '66)

66 (INTRO. 10-7-65)

(ONLY 4547 BLT.)

232 CID 6 (145 HP @ 4300 RPM) OR 287 CID V8 (198 HP @ 4700 RPM) 2 OTHER V8s AVAIL., TO 327 CID)

SALE $2601.
REG. $3051.

SOME WITH CORRUGATED ROCKER PANEL TRIM

NOW A PART OF AMBASSADOR LINE FOR '67.

new 118" WB

ROUND MEDALLION REMOVED FROM REAR DECK

67 (ONLY 2545 BLT.)

6 CYL. has new 155 HP @ 4400 RPM; STD. V8 with new 290 CID (200 HP @ 4600 RPM) OR 2 new 343 CID V8s (TO 280 HP @ 4800 RPM)

$3315. (WEST COAST)

new SMOOTHER BODY SIDES, RECTANGULAR GAS FILLER DOOR

RALLY LTS. IN GRILLE

THE FINAL MARLIN

BASE PRICE RAISED TO $2963.

REPLACED 1968 BY REBEL and JAVELIN

AMERICAN MOTORS

REBEL

(1967 TO 1970)

(FORMER CLASSIC MODEL)

DASH

$2863. SED.

770 WAG.

$3049.

REBEL SST HARDTOP

114" WB
232 CID 6 (145 HP)
OR
STD. 290 CID V8 (200 HP)

67

"550" FROM $2739.

UN-GRAINED

REBEL WAGONS $3155.

new SST "INTAKE" AHEAD OF REAR WHEELS (ALSO IN '68)

SST CONVERT.

$2872.

1967

67½

MARINER WAGON

ANCHOR DECOR ON UPHOL.

VARIOUS UNIQUE WOODGRAINS and INTERIORS FOR REBEL WAGONS (SPRING, '67)

BLEACHED TEAKWOOD PLANK WOODGRAIN EFFECTS ON BODY

WITH "TYPHOON" V8 ENGINE

AMERICAN MOTORS REBEL

550 (6 CYL.) WAGON

(550, SST ARE ONLY AMC CONVERTS. STILL AVAIL.)

SST (290 CID Typhoon V8)

6-CYL. 770

DASH

SST

(INTRO. 9-26-67)

68

SQUARE, RECESSED DOOR HANDLES (new)

SST "INTAKE" AHEAD OF REAR WHEELS

new SAFETY SIDE LIGHTS

68 SST

new 3-PIECE TAIL-LIGHTS

SST (OWN GRILLE)

ENGINES = 232 CID 6 (145 or 155 HP); 290 CID V8 (200 HP); 343 CID V8 (235 or 280 HP)

An intermediate-sized car for the price of a compact.

114" WB

Rebel $2,484 (SALE)

69

new GRILLE new TAIL-LTS., WIDER TRACK

(INTRO. 10-1-68)

(REG. $2944.)

BY GENERAL MOTORS GM

BUICK

(full-sized)

(SINCE 1903)

(STARTS OCT., 1945)

46

PRICE RANGE
$1391. - $2594.

SERIES		
40	SPECIAL	121" WB
50	SUPER	124"
70	ROADMASTER	129"

SPECIAL (RARE) HAS MORE HORIZ.
CHROME FENDER STRIPS. (SEE 1948)
ALSO.)

CLOCK

DASH
(WOODGRAIN EFFECTS ON
ROADMASTER)

CLOSE-UP OF
SPEEDOMETER

RE-SETTABLE
TRIP ODOMETER
(TO 999.9 MILES)

STRAIGHT-8 O.H.V.
ENGINES (SINCE 1931)
  SPECIAL, SUPER = 248 CID (110 HP)
  ROADMASTER = 320.2 CID (144 HP)

$2594.

ESTATE WGN. (SUPER)

748 BLT.

$1741.

34 425 BLT.

SUPER

$2046.

5987 BLT.

new GRILLE
(FEWER VERTICAL
PCS. THAN '42)

SUPER

77,724 BLT.

$1822.

6.50 x 16" TIRES
(7.00 x 15" ON ROADMASTER)

$2110.

20,864 BLT.

REAR VIEW

TOTAL 1946
BUICK PROD.:
153,627

ROADMASTER

MODEL 51

SUPER SEDAN IS BEST-SELLING MODEL (83,576 BLT.)

MODEL 56-C

$2333.

1929.

Super

2940.

MODEL 59

(28,297 BLT.)

CONVERT. has POWER-OPERATED TOP, FRONT SEAT and SIDE WINDOWS!

47

(272,827 BLT.)

new GRILLE has EMBLEM AT TOP, PLUS "BUICK EIGHT" LETTERING

WAGON (2036 BLT.)

POSTWAR "BOMBSIGHT" (ROCKET-THRU-RING) HOOD ORNAMENT (SINCE '46)

1947 PRICE RANGE: $1497. TO $3030.

($1611. TO 3249. PRICE RANGE LATER IN '47 SEASON)

ROADMASTER SEDAN (47,152 BLT.) $2232.

MODEL 71

note 1947 MODEL NAME AT UPPER CENTER of BUMPER GUARD

(213,599 BLT.)

SPEC.-SU.-RDMSTR. PRICES: (1948) $1735. TO $3433.

SUPER MODEL 56-S SEDANET (46,917 BLT.)

'48 SUPER and ROADMASTER NAMES ALSO APPEAR ON FRONT FENDERS.

MODEL 41 SEDAN $1809. (14,051 BLT.)

SPECIAL (CONTINUES WITH 46-48 STYLING INTO 1949)

MODEL 76-C

$1987.

48

(11,503 BLT.) CVT.

ROADMASTER $2837.

SEDAN $2418.

OPTIONAL: new

Dynaflow AUTO. TRANS.

(47,569 BLT.)

MODEL NAME

BUICK

Super MODEL 51 SEDAN (136,423 BLT.) $2157.

has 3 "PORTHOLES"

1949 HORSEPOWER RATINGS
SPECIAL   110
SUPER   115 or 120
ROADMASTER   150

248 or 320 CID STRAIGHT-8

CVT. TOP REAR DETAIL →

1949 PRICE RANGE: $1787. TO $3734.

49

TOTALLY RESTYLED (EXCEPT SPECIAL)

BACK SEAT (W. FOLD-DOWN ARM REST)

ROADMASTER
with Dynaflow Drive

ROADMASTER

has 4 "PORTHOLES"

MODEL 71 SEDAN

(55,242 BLT.)

$2735.

1949

new DASH

SEDANET

MODEL 79 STEEL-BODIED WAGON

$3734. →   (653 BLT.)

$3203.   (4343 BLT.)

"RIVIERA" (new H/T)   4343 BLT.

(EARLY MODEL WITH ALL-HORIZONTAL SIDE TRIM)

LATER 1949 CVTS. and RIVIERAS have new "SWEEP-SPEAR" SIDE TRIM (AS ILLUSTRATED)

$3150.

CVT. (8244 BLT.)   **TOTAL PRODUCTION: 324,276**

18

BUICK

$1856.

SPECIAL INTRODUCED EARLY (IN AUG., '49)

115 HP SPECIAL

121½" WB EXCEPT ON "52" SUPER SEDAN (125½") and ON 126½" and 130¼" RDMSTRS.

$1941. UP

SPECIAL RETAINS 2-PC. WINDSHIELD

STARTLING new "BUMPER-GRILLE"

new EASY-EYE TINTED GLASS IS OPTIONAL

MODEL YEAR PRODUCTION:
(552,827 CAL. YR.) 588,439
256,514 SPECIALS; 253,352 SUPERS;
78,573 ROADMASTERS

50 RESTYLED

ESTATE WAGON

THE ESTATE WAGON is yours on either SUPER or ROADMASTER chassis. Three power ranges to choose from.

128 HP **Super**

DASH (SPC.)

SUPER (2480 BLT.)
ROADMASTER (420 BLT.)

(114,745 BLT.) 4-DR. SEDAN

(54,512 BLT.)

BACK SEAT (SEDAN)

MODEL 15-R

152 HP **ROADMASTER**

MODEL 72

(66,762 BLT.) RIVIERA H/T

$2738.

130¼"WB ON "RIVIERA SEDAN"

ELONGATED "PORTHOLES"

(4 ON RDMSTR.)

19

BUICK

SPECIAL   SPECIAL DE LUXE   ESTATE WAGON   120 TO 128 HP   ('51)

(999 LOW-PRICED '51~'52 SPECIALS had 2-PC. WINDSHIELDS. = RARE!)

SUPER 128 HP

DASH ('52) (NOW SINGLE SPEEDO. DIAL IN FRONT OF STEER. WHEEL)

ROADMASTER   152 HP   170 HP

('51) ROADMASTER ('52)

('51)

PRODUCTION	1951	592,511
(MODEL		404,695
YEAR	1952	367,760
and CALENDAR YEAR)		321,048

('51)

51-52

← 1951 DASH (ABOVE) had **2** DIALS OF GAUGES IN FRONT OF STEERING WHEEL.

SUPER

'52 has BROADER HUBCAP MARGINS, FULL-HEIGHT VERTICAL BUMPER GUARDS, NO REAR FENDER CHROME STRIPS.

SPECIAL BUSINESS/CLUB COUPE (RARE!)   ('51)

Buick Eight

'51 BUMPER GUARDS DO NOT RUN DOWN FRONT OF BUMPER, BUT REST ON TOP.

Equipment, accessories, trim and models are subject to change without notice.

BUICK

$2255.

100,312 BLT.

FINAL STRAIGHT-8 IN SPECIAL

BUICK

SPECIAL RIVIERA 2-DR H/T

SPECIAL (125 HP @ 3800 RPM)

CVT.

$2553.

$2295.

58,780 BLT.

121½ W.B. ON ALL EXCEPT 125" WB RIVIERA SEDAN

(4282 BLT.)

SPORT WIRE WH. AVAIL.

LOWER PORTION OF DASH FINISHED IN BRIGHT MACHINE-TURNED METAL

1953 PRODUCTION: 485,353

1903-1953

53

BUICK'S GOLDEN ANNIVERSARY

$2696. "RIVIERA" SED.

SUPER

new V8

322 CID ENGINE IN ALL BUT "SPECIAL"

LIMITED-PRODUCTION "SKYLARK" CVT.

$5000.

164-170 HP (SUPER)

1690 BLT.

ROADMASTER RIVIERA

$3358.

$4031.

ONLY 670 BLT.

ROADMASTER

(188 HP @ 4000)

$3506.

BUICK

46-R

$2305.

71,186 BLT.

66-R

DASH

$3163.   ALL V8s

143 OR 150 HP
SPECIAL
(40 SERIES)
122" WB

1650 BLT.
1563 BLT.

49

195 OR 200 HP
CENTURY
122" WB

$3470.

CENTURY (60 SERIES)
RETURNS (1ST SINCE 1942)

69

45,710 BLT. $2534.

100-M

(100 SERIES)
SKYLARK *
$4483.

54

RESTYLED

ONLY 836 BLT. '54

* AFTER 1954, NO MORE
SKYLARKS UNTIL NAME
RETURNS IN 1961 AS PART OF
NEW COMPACT "SPECIAL" SERIES.)

264 OR 322 CID V8s
(THROUGH '55)
(264 CID IN
SPECIAL ONLY) $3269.

72-R
26,862 BLT.

NEW "PANORAMIC"
WRAP-AROUND
WINDSHIELDS (ALL MODELS)

1954 PRODUCTION:
442,903

$2626.   73,531 BLT.

56-R

127" WB   SUPER
(50 SERIES)

(70 SERIES)   ROADMASTER

$2964.

3343 BLT. 56-C

177 OR 182 HP
SUPER

3305 BLT.

ROADMASTER
200 HP @ 4100 RPM

127" WB

76-R $3373.

BUICK

SPECIAL

CENTURY

SUPER

150 TO 236 HP
$3453.

ROADMASTER

NOW 4 PORTHOLES ON ALL EXCEPT SPECIAL.

$2876.

new GRILLE and TAILLIGHTS (4-DR. HARDTOPS NOW AVAIL.)

55

DASH (SUPER)

new MESH GRILLE

1955 PRODUCTION : 737,035

MODEL 49 "SPECIAL" WAGON #3,770 BLT.

$2775.   *new TEARDROP-SHAPED PORTS*

CENTURY   "RIVIERA" 4-DR H/T   63   20,891 BLT.

SPECIAL (220 HP @ 4400 RPM)

$3025.

ALL MODELS NOW HAVE 322 CID V8s; SUPER, CENTURY and ROADMASTER HAVE 255 HP @ 4400 RPM

56

new GRILLE new TAILLIGHTS

76-R 12,490 BLT.

$3591.

$3204.

SUPER

29,540 BLT.

"RIVIERA" 2-DR. HARDTOPS

ROADMASTER

122 OR 127" WB (SINCE '54)

1956 PROD. : 635,158

new V-GRILLE with FINE HORIZONTAL PCS.

MODEL 56-R

DASH

BUICK

SPECIAL

122" WB
SPECIAL

CVT.
46-C
8505 BLT.

$2987.

57
new GRILLE

$3047.

49
7013 BLT.

1957 PRODUCTION :
404,049
26,589 BLT.

66-C

REAR
QUARTER WINDOW
DETAILS ON
CONVERTIBLE (WITH
TOP UP) (CENTURY)

$3270.

63
"RIVIERA"
4-DR. H/T

MODEL NAME
ABOVE DIP IN
SIDE CHROME
TRIM (EXCEPT
ON SPECIAL,
WHICH HAS NO
NAME HERE, OR ON
SUPER (with 3
CURVED CHROME
PCS. HERE.)

$3354.

CENTURY
122" WB

$3706.

66-R
17,029
BLT.

69

ALL
WITH
new
364 CID
ENGINE
(CONT'D.
IN
SPECIAL
and
LE SABRE
MODELS
THROUGH
1961)

new
CENTURY
CABALLERO
WAGON
(H/T
STYLE)

7.60 X 15

10,186 BLT.

new 127½" WB
SUPER

CVT.
56-C

56-R
$3536.

PILLARED
2-DOOR
SEDAN ONLY
AVAIL. IN
SPECIAL and
CENTURY (ONLY 2 BLT.)
SERIES (MODELS
48 OR 68)
(23,180 SPECIAL 2-DRS.)

DASH

4 DR. H/T

ROADMASTER
new 127½" WB
SERIES
70 OR 75

note
UNUSUAL
REAR
DOOR
TREATMENT
ON THIS MODEL

2-DR. H/T

300 HP @ 4600 RPM
(EXCEPT SPECIAL,
WHICH HAS 250 HP @
4400 RPM
(THROUGH '58)

BUICK

GO

SPECIAL

46-C

new BLOCK-STYLE GRILLE

HEAVY USE of CHROME TRIM

58
RESTYLED

SIDE "PORTHOLES"
DISCONTINUED
(UNTIL '60)

250-HP
SPECIAL

$2820.

FOR MODEL 43
"RIVIERA" 4-DR. H/T
31,921 BLT.

1958 PRODUCTION: 240,659

$3041.
46-C CVT.
(5502 BLT.)

$3154.
49 WAGON
3663 BLT.

THE
AIR BORN   B-58 BUICK

(ONLY 2-DR., 4-DR. H/Ts in SUPER SERIES)

DASH (RDMSTR.)

300 HP IN CENTURY,
SUPER,
RDMSTR.,
LIMITED

SPECIAL

$2636.
48 2-DR. SEDAN 11,566
(PILLARED)   BLT.

$3436.

63   CENTURY

15,171 BLT.

$4557.
75-R
"RIVIERA"
2-DR. H/T
2368
BLT.

DIAGONAL CHROME
STRIPS ON
LIMITED
R.R. FENDERS

75

10,505 BLT.

ROADMASTER

$4667.

$5112.

750   5571 BLT.

new
LIMITED
(NAME
REVIVED)

ONLY 839 BLT.   **$5125.**

CVT.
756

BUICK

ALL-NEW MODEL NAMES FOR 1959: LE SABRE, INVICTA, ELECTRA, ELECTRA 225

LE SABRE

4411 2-DR. 13,492 BLT.

123" WB

$2740.

$3129.

LE SABRE has 364 CID V8, 250 HP @ 4400 RPM

PACE CAR AT INDIANAPOLIS 500 RACE

DASH

MODEL 4419 4 DR. SED. 51,379 BLT.

$2804.

new CANTED HEADLIGHTS

INVICTA

The most spirited Buick

123" WB

$3515.

4639 4-DR. HARDTOP 20,156 BLT.

INVICTAS and ELECTRAS have 401 CID V8, 325 HP @ 4400 RPM

MODEL 4619 INVICTA SEDAN

(10,566 BLT.)

$3357.

59

TOTALLY RESTYLED

1959 PRODUCTION: 284,248
$2740. ~ 4300.
PRICE RANGE

new SQUARED-OFF ROOFLINE ON 4-DR. H/T

(10,491 BLT.)

$4300.

ELECTRA, ELECTRA 225

The most luxurious Buick

$4300.

10.5 COMPRESSION IN 1959

TOP-OF-LINE ELECTRA 225 IS ILLUSTRATED.

BUICK

$3145.

$2915.

LE SABRE
has 210, 235, 250 or 300 HP

364 CID = LE S.
401 CID = OTHERS
(SAME CHOICES IN '61)

60

A RETURN
TO VARIOUS "PORTHOLE" TYPE SIDE
DECORATIONS AS USED ON
1949 - 1956 BUICKS

new "Mirromagic" INSTRUMENT CLUSTER LETS DRIVER SEE GAUGES IN A MIRROR THAT CAN BE TILTED TO SUIT DRIVER'S OWN EYE LEVEL.

INVICTA WAGON
$3841. UP

INVICTA

1960 PRODUCTION : 253,807

123" WB = LE S., INVICTA
126.3" WB = EL., EL. 225

HEADLTS. PLACED HORIZONTALLY,
new GRILLE with CONCAVE VERTICAL PIECES and new 3-SHIELD BADGE

INVICTA
$3515.

1960 Buick Invicta 4-Door Hardtop in Magic-Mirror Tahiti Beige and Cordovan

$3357.

INVICTA and ELECTRAS have 325 HP @ 4400 RPM

$2756. ~ 4300. PRICE RANGE

4 PORTHOLES ON ELECTRAS, 3 ON OTHERS

H/T $3818. ELECTRA

ELECTRA 225

$4300.

BUICK

$2993.~4350.
PRICE RANGE

(new ROOFLINE)
$3228.

LE SABRE

123" WB
250 HP
LE SABRE

4439 4-DR. H/T
37,790 BLT.
(MOST POPULAR
1961 BUICK)

SPECIAL-SIZE
BUICK SPECIAL
new COMPACT SERIES,
STARTING 1961
SEE BUICK SPECIAL

SCARCEST
1961 BUICK
IS MODEL
4445
3-SEAT
WAGON
(2423 BLT.)
(ITS COMPACT
"SPECIAL" 3-SEAT
COUNTERPART SAW
ONLY 798 BLT.)

4639 4-DR. H/T
18,398 BLT.

INVICTA

$3515.

123" WB   325 HP

(TOTALLY
RESTYLED)
1961 PRODUCTION:
(FULL-SIZED MODELS)

61
191,392

MODEL 4829 4-DR.
"RIVIERA" H/T

$4350.

126" WB   ELECTRA 225   13,719 BLT.

STD. ELECTRA
SERIES INCLUDES
4719 SEDAN
(13,818 BLT.) $3825.
4737 2-DR. H/T
(4250 BLT.) $3818.
4739 4-DR. H/T
(8978 BLT.) $3932.

THORNWOOD

ELECTRA 225
CONVT. ———→
MODEL 4867
7158 BLT.
$4192.

28

BUICK

2-DR.

4-DR. H/T

4 DR. H/T

LE SABRE

$3567.

DASH

GAUGES HOODED, TO PREVENT WINDSHIELD GLARE

62

401 CID V8s (in all FULL-SIZED MODELS)

FULL-SIZED BUICK PRODUCTION: 256,766

ADVANCED THRUST

LE S. 265 HP @ 4400
INV. 280 HP @ 4400
ELEC. 325 HP @ 4400

H/T

INVICTA

4-DR. H/T

CVT.

$3815.

INVICTA ESTATE WAGON

$4034. (6-PASS.)

ELECTRA 225

4-DR. H/T

EL. 225 CVT.

SEDAN

H/T

ELECTRA has 126" WB (OTHERS 123")

Close-up of Wildcat! shows you new medallion and unique fabric overlay (available in black or white).

new WILDCAT!

325 HP

H/T

$4125.

BUICK

401 CID V8s IN ALL MODELS

LESABRE 265 or 280 HP

SEDAN

$3004. (64,995 BLT.)

1963 PRODUCTION: 327,123 (FULL-SIZED)

DASH

ROOF RACK

INVICTA (FINAL YR.) STATION WAGON IS THE ONLY '63 INVICTA

INVICTA WAGON $3969.

(3495 BLT.)

$4167. (WEST)

new V-SHAPED FRONT

63

325 HP (EXCEPT LESABRE)

$3961

4-DR. H/T (17,519 BLT.) $3871.

WILDCAT CVT. (6021 BLT.)

new!

WILDCAT has ITS OWN UNIQUE GRILLE, BUT TAILLIGHTS LIKE OTHER FULL-SIZED 1963 BUICKS.

WILDCAT

2-DR. H/T (12,185 BLT.)

$4047. (WEST COAST)

(6347 BLT.)

$3849. BASE PRICE

$4365.

(14,268 BLT.)

$4051.

Electra 225

$4141. (WEST COAST)

$2869.~4365. PRICE RANGE

ELECTRA TAIL-LIGHT DETAIL

(19,714 BLT.) $4186.

BUICK

4467 CVT.

$3314.

$3635.

ESTATE WAGON
4645 3-SEAT

6685 BLT.

4447 H/T

LE SABRE

400.3 BLT.

64

(257,438 BLT.)

$3061.

24,177 BLT.

$3164.

WILDCAT
325 HP

THE WILDCAT CONVERTIBLE
4667

4669
THE WILDCAT 4-DOOR SEDAN

7850 BLT.

WILDCAT (CLOSE-UP)

$3267.

DASH

ELECTRA 225

THE ELECTRA 225 4-DOOR HARDTOP

(ABOVE) WILDCAT
MODEL 4647 H/T
22,893 BLT.

LE SABRE 4400 HAS
new 300 cid V8 (210
or 250 HP.) LARGER
MODELS USE 401 cid V8
(325 HP.) as STD. EQUIPMENT.

THE ELECTRA 225 CONVERTIBLE

$4357.

new 425 cid V8
(340 OR 360 HP) is
OPTIONAL (LE SABRE
WAGONS, WILDCAT,
ELECTRA 225)

BUICK

LE SABRE CUSTOM CVT.
(6543 BLT.)

LE SABRE

$3030. UP

$3325.
123" WB
300 CID
210 HP
8.15 × 15 TIRES

LE SABRE 400

H/T
(VINYL TOP)
(21,049 BLT.)
(CUSTOM)
$3100.

$2948.
TO 4440.
PRICE RANGE

123 OR 126" WB
(SINCE '59)

65
new GRILLES

$3345.
TO
$4530.
PRICE RANGE
(WEST COAST)

ELECTRA 225 4 DR. H/T
(12,842 BLT.)
$4206.

(MORE
ELECTRA 225s
AT BOTTOM OF
PAGE)

TOTAL 1965 PRODUCTION: 653,838
FULL-SIZED (EXC. RIVIERA): 368,973

4 - DR.
H/T

8.45 × 15
TIRES
126" WB
401 CID
(325 HP)
425 CID V8 AVAIL.
(340 OR 360 HP)

WILDCAT
(STD., DLX.
and CUSTOM MODELS)

H/T

$3286.
UP

SEE ALSO:
RIVIERA

(8505 BLT.) CVT.

$4440.
ELECTRA 225
H/T
$4206. UP

8.85 × 15 TIRES

H/T

32

BUICK | $2942. ~ 4378. PRICE RANGE | 315,639 BLT. (FULL-SIZED, EXC. RIVIERA)

(SEPARATE SECTION FOR BUICK COMPACT CARS:
SPECIAL, SKYLARK, **1966 Buick.**
SPORT WAGON.) **The tuned car.**

LE SABRE (BELOW) 123" WB

WILDCAT (325 HP) (GS - 340 HP) 126" WB

$3547.

WILD. CUSTOM $4037.
(WEST COAST) (10,800 BLT.)

WILDCAT has UNIQUE GRILLE

Electra 225
(10,792 BLT.) 4 DR. H/T

66
ON COWL SIDES,

(CUSTOM MODELS IN EACH SERIES ARE HIGHER-PRICED.)

Electra 225 has 4 CHROME SEGMENTS,
LE SABRE has 3, and
WILDCAT has NONE.

$4153.
H/T (13,760 BLT.)

LE SABRE

220 HP $3142.

$3172. CUST. H/T $3560.
(22,666 BLT.)

4 DR. H/T (17,464 BLT.)

(10,585 BLT.)

DASH

(INTRO. 9-29-66)

67
new SLANTING SCULPTURED LINES RUNNING LENGTH OF BODY SIDES

WILDCAT $3382.

430 CID. 360 HP V8 (THROUGH '69)
336,366 BLT. (FULL-SIZED, EXC. RIVIERA)

Electra 225
H/T

(6845 BLT.)
430 CID V8 $4075.

ELECTRA 225 LIMITED

33

BUICK

Wouldn't you really rather have a Buick?

4 DR. H/T
(18,058 BLT.)
$3281.

(ABOVE)
LE SABRE 123" WB

(SHOWN WITH and WITHOUT VINYL TOP)

8.45 x 15" TIRES

WILDCAT 126" WB
CUSTOM H/T
(11,276 BLT.)

ENGINES:
350 CID V8
(230 OR 280 HP)
430 CID V8
(360 HP)

new
GRILLES

68

(INTRO.
9-21-67)

$3742.

ELECTRA 225
LIMITED 126" WB

PRODUCTION (FULL SIZED) = 384,575
ALL 1968 BUICKS = 651,823

$4597.
(WEST COAST)
8.85 x 15"
TIRES

INTERIOR

(ELEC. 225 LTD.)

126" WB
**ELECTRA
225**
CUSTOM SEDAN
(10,910 BLT.)
$4415.

$3141.~4541. PRICE RANGE

34

BUICK

No wonder Buick owners keep selling Buicks for us.
Wouldn't you really rather have a Buick?

350 CID
V8
230 OR 280
HP

LE SABRE

4-DR. H/T FROM $3740.
$3356. (WEST COAST)

69

(INTRO. 9-26-68)

LE SABRE
GRILLE has 5
HORIZONTAL PCS.;
(ILLUSTR. WILDCAT
GRILLE has
JUST ONE.)

WILDCAT

$3596.

430 CID V8
(360 HP)

WILDCAT
COWL
DETAILS

FORMAL
ROOFLINE

ELECTRA 225

(ELECTRA
GRILLE AT
LOWER
LEFT)

430 CID
V8
(360 HP)

TOTAL 1969 PROD.: (ALL MODELS) 665,422

$3216. ~ 4643. PRICE RANGE

EL.225 H/Ts FROM
$4323.

2-DR H/T, 4-DR H/T,
4-DR. SEDAN and CONVERTIBLE
IN EACH MODEL SERIES.

35

(INTRO. WED., OCT. 5, 1960, AS SEPARATE COMPACT SERIES of BUICK)

SPECIAL-SIZE

BUICKSPECIAL

THE BEST OF BOTH WORLDS

(and SKYLARK)

NEW!

BUICK'S REVOLUTIONARY ALUMINUM V-8. This hot 155 HP Fireball V-8 weighs just 318 pounds for a .487 horsepower to weight ratio — highest in the industry!

$2384.

3 VIEWS OF SPECIAL (SEDAN)

WAGON *has* 1-PIECE SWING-UP REAR DOOR

61

PRICES START AT
$2659. (WEST) (STD. CPE

SPECIAL WAGON

$2681. UP   112" WHEELBASE $3091. UP (WEST)

THE CLEAN LOOK of action

$2330. ~ 28/6.   PRODUCTION:
PRICE RANGE (BASE)   99,893

BUICK skylark

SKYLARK IS *new* LUXURY 185-HP MODEL of SPECIAL SPT. CPE.

SKYLARK IS AVAILABLE IN TWO-TONE OR SOLID COLORS (AS ILLUSTRATED)

$2949. (WEST)

112" WB and 6.50 x 13 TIRES (THROUGH '63)

note THAT SKYLARK *has* OWN REAR STYLING

36

BUICK SPECIAL

$2304.

SPECIAL
2-DR. CPE.
(19,135 BLT.)

$2358. 4-DR. SEDAN
(23,249 BLT.)

CONVERTIBLE
$2587.

(7918 BLT.)

SPECIAL

(2814 BLT.)

SPECIAL
DLX.

3-SEAT
WAGON
2736.

$2890.

15.5 OR
185-HP V8 (215 CID)
OR new V6 (198 CID, 135 HP)
ENGINE

WAGON
(10,380 BLT.)

$2304. ~
3012.

PRICE
RANGE

62

$2879.

$3012.

CVT.
(8913
BLT.)

SKYLARK

CVT.
(8332
BLT.)

SPECIAL DE LUXE

(34,060 BLT.)

H/T
$2787.

SPECIAL
2-DR. SPORT COUPE

6.50 x 13 TIRES
(SINCE '61)

$2818.
WAGON (8771
BLT.)

SPECIAL
DE LUXE

SEDAN

$2682.

(WEST COAST)

135, 155 OR
200 HP

PRODUCTION:
159,126 ('62); 152,226 ('63)

(37,695
BLT.) $2521.

$2309. ~ 3011. PRICE RANGE

63

$2857.

(10,212
BLT.) $3011.

(32,109 BLT.) H/T

SKYLARK

CVT.

BUICK SPECIAL

DELUXE 4-DOOR SEDAN

$2689.

4035
Special 2-seat Station Wagon
6270 BLT.

SPEC.
$2397.

$2490.
SPEC. DLX.

31,742 BLT.

4169

17,983 BLT.

4069

4337

$2669.

MODEL 4369 SEDAN
19,635 BLT.

SKYLARK ←

SKYLARK
has HEAVIER
SIDE TRIM WHICH
ENCOMPASSES THE
3 RECTANGULAR
"PORTHOLE" PCS.
IN COWL SECTION.

$2680.
SPORT COUPE
42,356 BLT.

64

new 115" WB
new 6.50 x 14 TIRES

4367 CONVERT.
10,225 BLT.

$2834.

new
SPORTS WAGON

SKYLARK

225 CID V6 (155 HP) OR
300 CID V8 (210 OR 250 HP)

PRODUCTION: 188,980

MODEL 4265
(9-PASS.)

$3124.*
(CUST.)

new
RAISED
PANORAMIC
ROOF WINDOWS,
AS ALSO
FOUND IN new OLDS
"VISTA-CRUISER"
WAGON.

REAR FENDER TRIM (WAGON)

Skylark

*$3562.
(WEST COAST)

INTERIOR VIEWS

MODELS 4255, 4265, 4355, 4365

**This is the new Buick Skylark Sports Wagon.
It has a raised roof so you can sit tall,
and a new kind of shaded glass so you can look
up and out, and a forward-facing third seat.**

120" WB (WAGON)

300 CID "WILDCAT" V8
IN WAGON

BUICK SPECIAL

PRODUCTION: 243,441

new 5-DIGIT MODEL NUMBERS

$2605. UP

44337 H/T

4549 BLT.

SKYLARK

43369
SPECIAL
SED.

$2397.

SPECIAL CVT.
43367 or 43467
(3357 BLT.) (3365 BLT.)

INDICATOR LIGHTS   HIGH BEAM

LEFT TURN   RIGHT TURN

65

new V-SHAPED FRONT ENDS, new GRILLE.

SKYLARK
GRAN SPORT
H/T 44437
47034 BLT. **$2751.**

INDICATOR LIGHTS   FUEL GAUGE   HEATER-AIR CONDITIONING CONTROLS   AIR CONDITIONER OUTLETS

CLOCK

LIGHTS WIPER   ACCESSORY   IGNITION   LIGHTER   RADIO   SPEAKER GRILLE

DASH

SKYLARK
GRAN SPORT
CPE.

11,877 BLT.
$2608. 44427

5309 BLT. 43427 (V8)
12 945 " 43327 (V6)

REAR

SPECIAL
2-DR.
COUPE **$2343.**
2414. (V8)

SKYROOF →
SPORTS WAG.

SKYLARK
with *new*
FULL-WIDTH
TAIL-
LIGHTS

H/T

1965 BUICK
Skylark
HARDTOP COUPE

BUICK COMPACTS

PRODUCTION: 216,709

(SPECIAL IS LOWEST-PRICED MODEL, FROM $2783.)
(GS) $2348. (WEST)

120" WB
Buick Sportwagon
FROM $3390. (WEST)

DASH (GS)
V
225 CID 6 (155 HP)
OR 300 CID V8
(210 HP)

115" WB

66

216,709 BLT. (COMPACT MODELS)

1966 Buick. The tuned car.

note COWL DECORATION

SKYLARK GS
(GRAN SPORT) (1835 BLT.)
$2956. H/T $3384. (WEST)

← GS WITH 401 CID, 325-HP WILDCAT V8 ENG.

194,355 BLT. 1967 COMPACT MODELS

$2411.~3340.
PRICE RANGE

67 new GRILLES

SPECIAL and SKYLARK 4-DOOR WAGONS FROM $2742.

$3167.

GS-400. (CVT.) (2140 BLT.)
$3563. (WEST)

new GS-340. (3692 BLT.)

$2845.

$2411. UP
SPECIALS (FROM $2844.,) (WEST)

GS-400
H/T (10,659 BLT.)
$3019.

SCARCEST MODEL (ONLY 894 BLT.) IS STANDARD SKYLARK V6 CPE.
$2665.

Skylark

H/T (41,084 BLT.)
$2798.

40

BUICK COMPACTS

9.25/8.55 x 14" TIRES
SPORTWAGON →

WAGON and SEDAN have 115" WB; CPE. and CVT. have 112" WB.

$3127. ↘

GS-400

(10,743 BLT.)

GS-400

"400" CID V8 (401)
(340 HP @ 5000 RPM)
3 SP. STICK SHIFT OR
3-4 SP. " " WITH
HURST SHIFTER OR
SUPER TURBINE AUTOMATIC

SPORT WAGON FR. $3711. (WEST)

Skylark (SKYLARK CUSTOM)

$2956.

VINYL COVERED TOP ↙

(44,143 BLT.)

2-DR. W.B. SHORTENED TO 112"

$3326. (WEST)

PLAIN TOP ↓↓

GS-400 ←

$3528 (WEST) ←

68 RESTYLED

GS 400 CVT. ALSO AVAIL.

7.75 x 14" TIRES

GS 400 DASH

(GS = GRAN SPORT)

GS-350
(H/T)

(GS-400 LOOKS SIMILAR)
350 CID V8
(280 HP @ 4600 RPM)

new 350 CID
(230 HP) V8
(STD. IN SKYLARK CUSTOM) RUNS ON REGULAR GAS.

(8317 BLT.)

$2926. $3295. (WEST)

'68 Buick. Now we're talking your language.

PROD.: 216,594

41

BUICK
COMPACTS

$2562. ~ 3621. PRICE RANGE

Sportwagon

WITH *new* "DUAL ACTION" TAILGATE →

8.55 x 14" TIRES

$4195. (WEST)

GS-350
350 CID V8 (280 HP)

"350"

350 CID V8 3 SIDE PORTS LIKE SPECIAL (DLX.)

GS-400 (BELOW)
400 CID V8 (340 HP)

$3181.

(6356 BLT.)

(4933 BLT.)

7.75 x 14" TIRES

$2980. $3810. (WEST)

7.75 x 14" W.S.W. TIRES

$3954. (WEST)

69
new GRILLES

SKYLARK
CUSTOM H/T
(35,639 BLT.)

$3739. (WEST)

"400 LETTERING

MODIFIED STAGE I GS-400 *has* "STAGE I" LETTERING INSTEAD, ALSO HI-LIFT CAM, 3.64 GEAR RATIO, ETC.

$3009.

7.75/ 8.25 x 14" TIRES

350 CID V8 STD. ON SKYLARK CUSTOM; AVAIL. ON OTHER SKYLARKS.

No wonder Buick owners keep selling Buicks for us.

Wouldn't you really rather have a Buick?

42

BUICK RIVIERA

luxury cars.

(STARTS 1963)

(by Buick)

401 CID
V8
325 HP @
4400 RPM

(40,000 BLT.)

117" WB

(DASH)

7.10 x 15
TIRES

63 MODEL 4747

America's bid for a great new international classic car

new 425 CID V8
(340 OR 360
HP)

(37,658 BLT.)

64 $4385.

MODEL 4747

'64 DASH

ADVENTURE IS A CAR CALLED RIVIERA AND IT'S A BUICK

(VINYL TOP ALSO AVAIL.)

new
TAIL-LIGHTS
IN BUMPER

65

MODEL 49447

new
CONCEALED
HEADLIGHTS

WIRE
WHEEL
OPTION

Wouldn't you really rather have a Buick?

$4408.   (34,586 BLT.)

2 ENGINES NOW AVAIL.: 401 CID V8 (325 HP)
OR 425 CID V8 (340 OR 360 HP)

BUICK

RIVIERA

new GRILLE

new 119" WB

1966 Buick. The tuned car.
425 CID V8 (360 HP)
8.45 × 15 TIRES
(45,348 BLT.)

$4424.

66

CONCEALED HEADLIGHTS

GS WHEEL COVER

"Riviera GS" ON COWL, JUST AHEAD OF DOOR

(INTRO. 9-29-66)
HAZARD FLASHERS at ALL CORNERS

67

TAIL LIGHTS

$4469.

new ENERGY-ABSORBING STEERING COLUMN

CENTER CONSOLE AVAIL.

(42,799 BLT.) $4557. (WEST COAST)

new 430 CID V8 (360 HP)
(THROUGH '70)

(INTRO. 9-21-67)

MODEL 49487
$4703.
$5245.
(WEST COAST)

VINYL TOP COVERING

PLAIN TOP

"RIVIERA" NAME ON COWL IS NOW IN BLOCK LETTERS.

68

(49,284 BLT.)

new SPLIT GRILLE

8.45 × 15" TIRES

| 1968 PRODUCTION | = 49,284 |
| 1969 PRODUCTION | = 52,872 |

new SIDE SAFETY LTS.

MODEL 49487

430 CID V8

$4701.

69

new 8.55 × 15" TIRES

$5321.
(WEST COAST)

(INTRO. 9-26-68)

1968

new GRILLE SPLIT, AS IN 1968.

new GRILLE has FINE VERTICAL PCS. and JUST 2 HEAVIER HORIZONTAL PCS.

Wouldn't you really rather have a Buick?

44

Cadillac Standard of the World

(SINCE 1902)

126" WB 61 (2200 BLT.) 6267 CVT. D $2556.

$2176. $2359. 62 ← 129" WB (1342 BLT.)

TOTAL 1946 PRODUCTION: 29,214

(14,900 BLT.)

FLEETWOOD 60-S

(60-S has 5 SLOPING CHROME STRIPS ON REAR QUARTER PANEL.)

46

(5700 BLT.) $3095.

75 (136" WB THROUGH '49) $4153. TO 4669.

AS IN 1942, A TOTAL OF 6 HORIZONTAL GRILLE MEMBERS, BUT WITH new RECTANGULAR GRILLE LTS.

FRONT VIEW

Cadillac Standard of the World

(5160 BLT.)

$2523. 61 126" WB

$2324. ↗

6289 SEDAN (25,834 BLT.) 62 129" WB

TOTAL OF 5 HORIZONTAL MEMBERS IN 1947 GRILLE.

new SMALL ROUND PARKING LIGHTS

47

TOTAL 1947 PROD.: 61,926

$2060. TO $4590. PRICE RANGE

62 6267 CVT. (6755 BLT.) $2902.

Cadillac NAME ON FENDERS IS NOW IN SCRIPT STYLE.

AS BEFORE, CHROME STRIPS IDENTIFY 60-S

133" WB

MODEL 6069 SEDAN

ALL MODELS HAVE L-HEAD 346 CID V8 (150 HP @ 3600 RPM) (SINCE 1941)

8500 BLT. $3195.

new HEAVY FLANGES ON 1947 HUBCAPS.

45

Cadillac
STANDARD OF THE WORLD

61 6/69 SEDAN (5081 BLT.) **$2833.**

PLAIN ROCKER PANEL ON "61."

new "FISHTAIL" REAR FENDER FINS

126" WB ON BOTH "61" and "62" SERIES

48

TOTALLY RESTYLED! (EXCEPT FOR SERIES "75")

REAR DETAILS

Cadillac

PRESENTS THE *New* STANDARD OF THE WORLD!

CLOSE VIEW OF ENAMELED CREST and CHROMED "V" ON FRONT OF HOOD

$2728. ~ 5199. PRICE RANGE

62 → $2912.

6207 (4764 BLT.)

has CHROME ROCKER PANEL STRIP

75

FLEETWOOD **60-S** (656I BLT.)

(60 SPECIAL SEDAN) 6069

"75" RETAINS OLD STYLING (THROUGH 1949) $4679. TO $5199.

$3820.

133" WB

60-S REAR FENDER HAS UNIQUE CHROME TRIM.

TOTAL 1948 PRODUCTION: 52,706

Cadillac
Standard of the World

61 2-DR.

$2840. (6409 BLT.)

49

IMPROVED V-8
ENGINE NOW HAS
OVERHEAD
VALVES.

160 HP
(THROUGH '51)
126" WB (61, 62)
GRILLE MODIFIED *

...The world's newest engine—for the world's finest car!

* EXCEPT ON 75,
1949 GRILLE HAS
ONE LESS HORIZONTAL
PIECE THAN 1948.
NEW CHROME
WRAP-AROUNDS
EXTEND GRILLE AT
EITHER END.

62
6269 SED.
(37,977 BLT.)

$3050.
($3103.,
WEST COAST)

6237 HARDTOP
(2150 BLT.)
$3497.

NEW "COUPE
DE VILLE"
(CADILLAC'S
FIRST H/T
(HARDTOP CVT.)
(PILLARLESS)

62

TOTAL 1949 PRODUCTION: 92,554

6267 CONVERTIBLE (8000 BLT.)

$3442.
($3549.,
WEST COAST)

(WEST COAST)
$5253.

133" WB
FLEETWOOD
60
SPECIAL

$3891. (11,399 BLT.)

75

"75" RETAINS
OLDER
STYLING.
136¼" WB

47

Cadillac
Standard of the World

6169 SEDAN
14,931 BLT.

61

AS ILLUSTRATED, NO REAR QUARTER WINDOWS ON 61 SEDAN

$2866. 122" WB

6237 D H/T
4507 BLT.

$3234.

$3523.

62

6219 SEDAN   41,890 BLT.
62
126" WB
new GRILLE

$2761 TO $4959
PRICE RANGE

TOTAL 1950 PRODUCTION: 103,857

6267 CONVERTIBLE   6986 BLT.
$3654.

new 1-PIECE WINDSHIELD

50

ALL MODELS RESTYLED (INCLUDING "75")

6019 SEDAN (BELOW) 13,755 BLT.

60-S 130" WB
NOW HAS SMALL LOUVRES HERE

$3797.

75
(new 146 3/4" WB) $4770. UP

CONVERTIBLE WITH CUSTOM INTERIOR (USUALLY LEATHER)

Cadillac
Standard of the World

61 6/69 4-DR. SEDAN

62/9 SEDAN
55,352 BLT.
$3528.

FINAL "61" MODEL
2300 4 DRS. BLT.
2400 2 DRS. BLT.
$2917.

62

6237-D "COUPE DE VILLE" H/T
10,241 BLT.
$3843.

51

FINAL YEAR FOR 160 HP

60 SPECIAL

DASH

18,631 BUILT $4142. MODEL 60/9

IGNITION-KEY STARTING

75 (ABOVE)
$5200. TO 5405.
146.8" W.B.,
INCLUDES:
8-PASS. SED. (1090 BLT.)
9- " BUSINESS SEDAN (30 BLT.)
8-PASS. IMPERIAL SEDAN (1085 BLT.)
ALSO 2960 (CHASSIS UNITS (157" W.B. COMMERCIAL USE)

new "WAFFLE" EXTENSIONS AT EITHER END OF 1951 GRILLE

TOTAL 1951 PROD.: 110,340

$2810. ~ 5405. PRICE RANGE

GOLDEN ANNIVERSARY

Cadillac
Standard of the World

BEAUTIFUL NEW INTERIORS IN ALL MODELS

"COUPE DE VILLE" H/T →

$4163. CVT. (6400 BLT.)

62 (NOW THE LOWEST-PRICED SERIES)

"V" INSIGNIA NOW COLORED GOLD, TO COMMEMORATE CADILLAC's 50TH ANNIVERSARY.

1902 52

STANDARD OF THE WORLD

52

"GOLDEN ANNIVERSARY" MODELS INCREASED BY 30 H.P.

THESE new DECORATIONS FOUND ON 1952 MODELS ONLY

8-PASS. "75" SEDAN

147" WB

(16,110 BLT.)

60-S

$4323.

NEW 190-HORSEPOWER ENGINE

★ NEW HYDRA-MATIC DRIVE

★ NEW FRONT AND REAR END APPEARANCE

★ NEW CADILLAC POWER STEERING

★ NEW DUAL EXHAUST SYSTEM

MODELS: 62/9 SEDAN = $3684. (42,625 BLT.)   75 SERIES: 7523 8-PASS. SEDAN $5428.
6237 COUPE   3587.   10,065   (1400 BLT.)
6237D H/T   4013.   11,165   7533 IMPERIAL SEDAN $5643.
6267 CVT.   4163.   6400   (800 BLT.)
60 SPL. 60/9 SEDAN   4323.   16,110

1952 PRODUCTION : 90,259

Cadillac Standard of the World

62

6267-S
EL DORADO
CVT. (new)
(has WRAP-AROUND
WINDSHIELD)
$7750.
(ONLY 532
BUILT)

26" WB   6237D 14,550 BLT.

60-S has MORE CHROME ALONG LOWER EDGE, PLUS THE CHARACTERISTIC VERTICAL STRIPS.

60-S $4341.
130" WB
6019 20,000 BLT.

$4144.
62 CVT.

53

DASH

6267 CVT.
8367 BLT.

new PARKING/ DIRECTIONAL LIGHTS

PRICE RANGE (ALL MODELS)
$3571.
TO
$7750.

(THROUGH 1955)
331 CID ENGINE
210 HP @ 4150 RPM

TOTAL 1953 PRODUCTION:
109,651

75 SERIES (146.75" WB)
BELOW: 7523 8-PASS. SEDAN-LIMO. (1435 BLT.)

$5408.

51

Cadillac
Standard of the World

62

new 129" WB
ON "62"

$4261.

CPE. DE VILLE
MODEL 6237.D
(17,170 BLT.)

(also, 17,460 new
STD. H/Ts at $3838.)

DASH
(CONVERTIBLE)
6267 (6310 BLT.)

$4404.

SPEEDOMETER
CLOSE-UP

A color picture was taken right through this E-Z-EYE Panoramic

$4863. 60-S

MODEL
60/9
(16,200
BLT.)

new 133" WB ON 60-S

new PANORAMIC
WINDSHIELD

75
LIMOUSINE
new 149.75" WB

54

RESTYLED
new 230 HP
@ 4400 RPM

SPECIAL
ELDORADO REAR
FENDER BRIGHTWORK

FRONT
VIEW

6267S ELDORADO
CVT. (2150 BLT.)

$4738.

TOTAL 1954
PRODUCTION:
96,680

"75" COMMERCIAL CHASSIS (#8680 S)
INCREASED FROM 157" TO 158" WB)
(1635 BLT.)
"75" 8-PASS. SEDAN $5875. (889 BLT.)*
" " " IMPERIAL SEDAN $6090.
(611 BLT.)*   * 149.75" WB

Cadillac
Standard of the World

129" WB
62

$3882.
TO
6402.
PRICE RANGE

HIGHLIGHT FEATURE
of CADILLAC and OLDSMOBILE for '55!

AUTRONIC -EYE

REAR VIEW

AUTOMATIC
LIGHT
'TROL

PADDED DASH DETAIL →

BRIGHT

DIM

BRIGHT

Automatically AT NIGHT!

(18,300 BLT.)

$4728.

← 60-S (133" WB)
MODEL 6019 SEDAN

new GRILLE with HEAVIER PIECES

'75 has VERTICAL CHROME STRIP RUNNING TO BOTTOM OF REAR FENDER.)

55

new
250 HP
@ 4600 RPM

TOTAL 1955 PRODUCTION:
140,777

ELDORADO MODEL 6267S

CONVERTIBLE
(3950 BLT.)

$6286.

270 HP @ 4800 RPM
(ELDORADO ONLY)

DORADO has OWN STYLE of new PAINTED FIN REAR FENDERS.

129" WB

2 VIEWS OF ELDORADO

IMPORTED, HANDCRAFTED LEATHER UPHOLSTERY in ELDORADO

Cadillac
Standard of the World

CONVERT.

COUPE DE VILLE
(24,086 BLT.)

$4766.

(8300 BLT.)

62

$4624.

(17,000 BLT.)

60-S

$5047.

PRICES
START AT
$4146.

(ACTUAL PHOTO)

LENGTH OFTEN
EXAGGERATED IN
ADVERTISING ART

$4753.

62
SEDAN
DE VILLE
new 4-DR. H/T
(41,732 BLT.)

REAR
DETAILS

new
365 CID
(THROUGH
'58)
285 HP @
4600 RPM

56

MORE and
FINER MEMBERS
IN GRILLE

Highway Beam Indicator
Hand Brake Warning Light   Odometer   Defroster Control Lever
Hydra-Matic Quadrant   Speedometer   Signal-Seeking, Pre-Selector Radio
Oil Pressure Warning Light   Generator Indicator Light   Map Light
Turn Signal Indicator   Turn Signal Indicator
Heater Control Lever   Gasoline Gauge   Lighter
Headlights

Hand Brake   Air Vent Control   Ash Tray
Temperature Indicator   Lighter   Lighter
Convertible Top   Air Vent Control   Ash Tray
Control Knob   Ignition Switch
Windshield Wipers and Washers   Odometer Reset Knob   Air Conditioner Control Panel

instrument panel and controls

TOTAL 1956 PRODUCTION : 154,577

$4146. ~ 6828.
PRICE RANGE

75
LIMOUSINE

(955 BLT.)

$6773.

SEVILLE

Eldorado
BY CADILLAC

(3900 BLT.)

BIARRITZ CVT.

$6501.

FOR EITHER MODEL
OF EL DORADO

(2150 BLT.)

305 H.P.
WITH
365 CID V8

AS IN 1955, ELDORADO has
OWN STYLE OF TAIL FINS.

Cadillac *Standard of the World*

$4677. TO **$13,074.**
PRICE RANGE (new EL.D. BRGH. IS COSTLIEST MODEL.)

$5614.

60-S

← REAR BRIGHTWORK PANELS ON 60-S

new SQUARED-OFF TAIL-FINS with LOW, ROUND TAIL-LIGHTS

new 300 HP (325 ON ELDORADO MODELS)

RESTYLED **57**

62

TOTAL 1957 PROD.: 146,841

new ELDORADO BROUGHAM (4 DR.) SEVILLE (2 DR.)

6267 CVT. (9000 BLT.) **$5293.**

new GRILLE

new 6267 S ELDORADO BIARRITZ CVT. (1800 BLT.) **$7286.**

4-DR. HARDTOPS REPLACE SEDANS IN 62, 60-S SÉRIES.

ALL 1958 CADILLACS HAVE SHARP, BACK-SLANTING TAIL FINS AS ON '57 ELDORADO BIARRITZ (ABOVE)

4-DR. H/T AVAIL. WITH or W/O EXTENDED DECK

LENGTH EXAGGERATED IN THIS ORIG. ADVERTISING ART

62

$5079.

129½ WB

6239 E EXTENDED-DECK 4 DR. H/T (20,952 BLT.)

new 310 HP

58

TOTAL PROD.: 121,778

FOUR HEADLIGHTS

6267 CVT. (7825 BLT.) **$5454.**

WIDTH EXAGGERATED IN THIS ORIGINAL ADVERTISING ART →

1958

new LOWER, BROADER GRILLE

MORE 1958 MODELS ON NEXT PAGE

$6232. _Cadillac_
STANDARD OF THE WORLD

60-S CONTINUES LOWER BRIGHTWORK PANELS ON REAR FENDERS. 133" WB

MODEL 6039
(12,900 BLT.)

AS IN 1957, EL DORADO BROUGHAM HAS ITS OWN UNIQUE FRONT END STYLING.
$13,074.
SERIES 70
MODEL 7059
(304 BLT.)

AIR SUSPENSION ON ELD. BROUGHAM, OPTIONAL ON OTHER 1958 CADILLACS.

A PRICE OF
$7500.
FOR SEVILLE OR BIARR.

(855 BLT.)

TIRES:
8.00 × 15 OR
8.20 × 15

Cadillac Eldorado

EL DORADO SEVILLE 6237S

(335 HP, 129½" WB ON EL DORADOS)

EL DORADO BIARRITZ 6267S

62

6239
4 DR. H/T
(STANDARD-LENGTH DECK)

(13,335 BLT.)

$4891.

20-GAL GAS TANK

58
(CONT'D.)

75 STYLED LIKE 62 SERIES, BUT W. 149 3/4" W.B. AVAIL. MODELS:
7523 9-PASS. SEDAN
(802 BLT.)
$8460.
7533 9-PASS. LIMO.
(730 BLT.)
$8675.
ALSO 156"-WB CHASSIS-UNITS (1915 BLT.)

note ROUNDED-DOWN REAR FENDER/DECK PANELS ONLY ON THESE 2 EL DORADO TYPES.

(815 BLT.)

Cadillac
STANDARD OF THE WORLD

ENORMOUS TAIL-FINS!

6237 H/T (21,947 BLT.)

new "DOUBLE-DECK" GRILLE

CLOSER VIEW OF TRADITIONAL "V" ON REAR DECK

DETAILS OF THE UNIQUE REAR END DESIGN IN 1959

GRILLE MOTIF IS ALSO CARRIED ON AT REAR

new 390 CID V8s
ON ALL EXCEPT ELDORADO
new 325 HP
(THROUGH '63)

new 130" WB
ON ALL EXCEPT
75 SERIES)

TOTALLY RESTYLED

59

DE VILLE NOW A SEPARATE SERIES FROM 62.

6267 "62" CONVT. (11,130 BLT.) $5455. (NO DE VILLE CVT.)

6339 "SWEEP-ROOF" SEDAN DE VILLE (12,308 BLT.)

2 DR. H/T $4892. (62)

DE VILLE

$5252. (DE V.)

NOTE THE ROOFLINE DIFFERENCES BETWEEN THESE 4-DOOR HARDTOPS

$5498.

1959 PRICES START AT $4892.

6329 "FLAT TOP" SEDAN DE VILLE (19,158 BLT.)

'59 HP FIGS. @ 4800 RPM

FLEETWD. 75

6733 LIMOUSINE $9748. (CONT'D.)

(149.87" WB)

(690 BLT.)

CADILLAC
STANDARD OF THE WORLD

59
(CONT'D.)

MODEL 6039 4-DR. H/T
(12,250 BLT.)

FLEETWOOD
60 SPECIAL
(note ITS
OWN UNIQUE
SIDE and FENDER
TRIM)

$6233.

note "FLEETWOOD" NAME
ON FRONT FENDER PANEL. (60-S)

DASH

345-HP
EL DORADO MODELS BELOW:

("ELDORADO" NAME
on FRONT FENDER
PANELS of
BIARRITZ and
SEVILLE ONLY.)
$7401. (EITHER
MODEL)

6929
BROUGHAM
(ONLY 99 BLT.)→

6467
BIARRITZ
(1320 BLT.)

EL DORADO
BROUGHAM STYLING
DIFFERS FROM
OTHER 1959
CADILLACS.
$13,075.

6437
SEVILLE
(975 BUILT)

2102 MODEL 6890 COMM. CHASSIS UNITS ALSO PRODUCED IN
"75" SERIES, ON A 156" WHEELBASE

TOTAL 1959 PRODUCTION: 142,272

Cadillac
STANDARD OF THE WORLD

$5455.

62

26.7 CVT.
14,000 BLT.)

new GRILLE

PRICES START AT
$4892.
FOR 2-DR. 62 H/T (ILLUSTR.)

6237
2-DR H/T
(19,978 BLT.)

2 DIFFERENT ROOFLINES STILL AVAIL. ON "62" OR "DE VILLE" 4-DR H/Ts

TAIL FINS REDUCED FOR 1960, IN STYLE OF '59 ELDORADO BROUGHAM.

60

SEDAN DE VILLE

$6233.

60-SPECIAL

(11,800 BLT.)

6039

TOTAL 1960 PRODUCTION:
142,184

FINAL H/T EL DORADOS (UNTIL 1967, AT WHICH TIME EL DORADO BECOMES A SPECIAL FRONT-WHEEL-DRIVE 2-DR. H/T.)

6929 EL DORADO BROUGHAM

(ONLY 101 BLT.)

note THAT THE EL DORADO BROUGHAM *has* SIDE TRIM DIFFERENT FROM THAT OF THE OTHER EL DORADO MODELS OF 1960.

6437

(1075 BLT.)
(SAME PRICES AS IN 1959, ON ALL 3 EL DORADOS)

EL DORADO SEVILLE

$7401.

467

EL DORADO BIARRITZ (1285 BLT.)
EL DO. CVT. CONT'D. THROUGH '66)

6733
FLEETWOOD 75
LIMOUSINE
(832 BLT.)

$9748.

9533. FLEETWOOD 75 9-PASS. SEDAN
ALSO AVAIL. (718 BLT.)

Cadillac

6229 4-DR. H/T
(26,216 SOLD)
$5080.

New 6399 "TOWN
SEDAN" 6 WINDOW
H/T IN DE VILLE
SERIES (3756 BLT.)

62

6237 2-DR H/T
(16,005 BLT.)

6267 CVT.
(15,500 BLT.)
$5455.

$4892.

62

DE VILLE
PRICE RANGE =
$5252-5498.

New
LOWER SIDE FIN,
TO BALANCE
EFFECT of
UPPER TAIL FIN

New
CONVEX
GRILLE

61

New 129½" WB
RESTYLED, SLIGHTLY DOWNSIZED and
LIGHTENED

TOTAL 1961 PRODUCTION:
138,379

$4892.
TO
$9748.
PRICE
RANGE
(62 TO 75)

CHROME BANDS NEAR
END OF REAR FENDER
IDENTIFY 60-S.

FLEETWOOD 60-S

$6233.

60

Cadillac STANDARD OF THE WORLD

DASH

RADIO, CLOCK DETAIL

62 new GRILLE and TAIL LTS.

"CONVERTIBLE-STYLE CREASE IN REAR ROOF LINE

COUPE DE VILLE

COUPE (2-DR. H/T)

SEDAN DE VILLE
6339
(27,378 BLT.)

DE VILLE

DE VILLE
6347
(25,675 BLT.)
$5385.

$5189.
(OR $5025.)
"62" 6247 2DR. H/T
(16,833 BLT.)

Fog Lamps

* ALSO PRICED AT $5631.

new "TOWN SED." MOVED TO 62 SERIES IN 1962, THEN DISCONT'D.

new 6389 "PARK AVE." 4 DR. H/T in DE VILLE SERIES (THROUGH 1963) (OPT.)

62

6267 CONVERTIBLE
(16,800 BLT.)
$5588.

$9937. OR
$10,100.

TOTAL 1962 PRODUCTION:
160,840

FLEETWD. 60-S

6733 FLEETWOOD 75 LIMO.
(904 BLT.)

BACK SEAT

REAR COMPARTMENT (75)

$6366. OR $6529.

Cadillac

TOP STYLING ON HARDTOP.

2-DR. H/T PRICES START AT $5026.

62

new CONVERTIBLE-STYLE HARDTOP ROOF LINES.

6267 CVT. (17,600 BLT.)

4W and 6 WINDOW 4-DR. HARDTOPS IN 62 and DE V. SERIES

$5590.

$9939.
FLEETWOOD 75 LIMOUSINE

6733 (795 BLT.)

PHOTO →
AD ILLUSTR. →

FLEETWOOD 75

FLEETWOOD 60-S
(14,000 BLT.)

6039

6723 9-PASS. SEDAN (680 BLT.)
$9724.

(EXAGGERATED LENGTH)

62

$6366.

DASH

"FLEETWOOD" ON 60-S FENDER

TOTAL 1963 PROD.: 163,174

REAR CLOSE-UP

$6608.

EL DORADO BIARRITZ
6367

63

(1825 BLT.)

HEAD-ON DETAIL OF LIGHTS IN RELATION TO GRILLE

new 340 HP (62 SERIES ONLY)

new GRILLE EMPHASIZES "DOUBLE-DECK" STYLING.

Cadillac
STANDARD OF THE WORLD

62

60-S

DASH AND INTERIOR VIEWS

new "COMFORT CONTROL" ←

64

ALL MODELS NOW HAVE 340 HP @ 4600 RPM and new 429 CID

new CONVEX GRILLE

PRICED FROM $5048. (62 H/T)

6267 CVT. (17,900 BLT.)
$5612.

FLEETWOOD 75
6723 9-PASS. SED.   6733 LIMO.
(617 BLT.)   (808 BLT.)
$9746. $9960.

62

60-S

75

"62" PRICES FROM
$5191.

Comfort Control combines heating and air conditioning in a single unit, the interior weather never changes. Even humidity is under perfect control. This system now available as an extra-cost option.

DE VILLE $5247.

new CALAIS (REPLACES 62 SERIES) $5247.

68357 H/T (43,345 BLT.)

CALAIS

129½" WB

68239 4-DR. H/T (13975 BLT.)

$5419.

DE VILLE

TOTAL 1965 PRODUCTION: 182,435

Cadillac

65

new TAIL-LIGHTS

new 4-DR. SEDANS AVAIL., IN ADDITION TO ONGOING 4-DR. H/Ts.

PRICE RANGE: $5224. TO $10,125.

new LARGE 1-PC. GRILLE

FLEETWOOD BROUGHAM

60 S (new 133" WB) 68069 SED. (18,100 BLT.) $6479.

new VERTICALLY-PLACED HEADLIGHTS

1965

DASH

COMFORT CONTROL

WIPER

Cadillac

Cadillac
STANDARD OF THE WORLD

RADIO DETAIL

64

Cadillac

DE VILLE
$5339.

COUPE DE VILLE H/T

FLEETWOOD 75
LIMO. or 9-PASS. SEDAN
$10,312. UP

DE VILLE CVT.

$5555.

(0,580 BLT.)

66

new GRILLE

$6695.

Interior (FLEETWOOD)

(13,630 BLT.)

FLEETWOOD BROUGHAM ↑
JOINS 60-S SEDAN

TOTAL 1966 PRODUCTION: 196,685

(INTRO. 10-6-66)

TOTAL 1967 PRODUCTION: 200,000

new GRILLE

67

FLEETWOOD BROUGHAM

$6739.

CALAIS   5215.

FRONT DETAIL

(2865 BLT.)

$6277.

New

ELDORADO H/T

(17,930 BLT.)

ELDORADO'S GRILLE DIFF. FROM OTHER MODELS

FLEETWOOD 75
(BELOW)

(has 8.20 x 15 TIRES)

SEDAN = $10,522.
LIMOUSINE = 10,733.
149.8" WB

75 INTERIOR
(7-PASS.)
IN
LIGHT GRAY
DEVONSHIRE
CLOTH

front to rear: DeVille Convertible; Fleetwood Eldorado; Coupe deVille; Fleetwood Brougham; Fleetwood Seventy-Five Sedan; Hardtop Sedan deVille; Sedan deVille; Fleetwood Sixty Special; Calais Coupe; Calais Hardtop Sedan; Fleetwood Seventy-Five Limousine.

FULL LINE OF 11 MODELS ILLUSTR.

new CONCEALED WINDSHIELD WIPERS

new 472 CID V8 375 HP (AVAIL. THROUGH '74, HP CUT TO 345 IN '71, TO 220 IN '72, TO 205 IN '74.)

Cadillac

SEDAN DE VILLE 4 DR. H/T (72,662 BLT.)

68

new GRILLES

H/T (AVAIL. IN CALAIS OR DEVILLE SERIES

CVT. (18,025 BLT.)

$5736.

TOTAL 1968 PROD.: 230,003

DETAILS OF ELDORADO'S RETRACTABLE HEADLIGHTS

(24,528 BLT.)

EL DORADO H/T

$6605.

COUPE DE VILLE H/T

$5721. (65,755 BLT.)

$7110.

FLEETWOOD BROUGHAM

DE VILLE $5905.

(16,445 BLT.)

DeVille Convertible

(17,300 BLT.)

69 new GRILLES

9.00 x 15" TIRES

Eldorado H/T

PLAIN TOP

$6711.

(23,333 BLT.)

VINYL TOP

(A TOTAL OF 23,333 1968 ELDORADO COUPES BLT.)

TOTAL 1969 PRODUCTION: 223,237

new ELD. GRILLE w. FINER PCS. new UNCONCEALED HEADLIGHTS

"The Hugger" **Camaro** (SINCE 1967) BY **CHEVROLET**

230 CID
6 CYL. OR V8 (327 OR 350 CID)
(140 TO 325 HP)
108" WB

(INTRO. 9-29-66)

$2466. UP

NEW 67

STOCK CAMARO 6 COUPE

SS CONVERTIBLE

SS-350 CPE. w. CONCEALED LTS.

1967 PRODUCTION:
6 = 58,808
V8 = 182,109

68 $2588. UP

1968 PRODUCTION:
6 = 50,937
V8 = 184,178

SS HAS DUAL ROWS OF SQUARE PORTS ATOP HOOD

new DASH

ENGINES
230 CID 6 (140 HP)
250 CID 6 (155 HP)
327 CID V8 (210 HP) (275 HP AVAIL.)
350 CID V8 (295 HP)
396 CID V8 (325 HP)

NOTE new FRONT and REAR SIDE (RECTANGULAR) SAFETY LIGHTS, AS REQUIRED BY LAW

Chevrolet Camaro (FINAL CONVERTIBLE AVAILABLE)

SS SPT. CPE. WITH RALLY SPORT EQUIP.

REAR FENDER

V-SHAPE and CRISS-CROSS PCS. IN new GRILLE → (BLK.)

69

(CONTINUES TO FEB., 1970)

140 HP 6 TO 396 CID 325 HP V8
(EARLY 327 CID, 210 HP V8 REPL. BY new STD. 307 CID, 200 HP)

1969 PRODUCTION:
6 = 65,008   V8 = 178,087

CHECKER (1922 TO 1982)

CHECKER MOTORS CORPORATION
Kalamazoo, Michigan

1947 TO 1955 STYLE →

1956 TO 1958 STYLE →

TAXIS, COMMERCIAL ONLY (THROUGH '58)

Checker Aerobus Limousine

CHRYSLER V8 ENGINE IN PRE-'64 AEROBUS

226 CID, 80 HP 6-CYL. CONTINENTAL ENGINE USED (UNTIL '63.) STARTING 1964, CHEVROLET 6 OR V8.

DASH ('68)

Checker Marathon Deluxe Limousine

NON-COMMERCIAL SEDANS and WAGONS NOW AVAIL. TO PUBLIC.

59 ON

NO YEARLY STYLE CHANGES.

120" WB

Checker Marathon 4-door sedan

Checker Marathon 4-door station wagon

SAFETY-BUMPERS (ENERGY-ABSORBING) ADDED IN MID-1970s.

INTERIOR ('69)

1965 ENGINES =
230 CID 6 (140 HP)
283 " V8 (195")
327 " " (250")
(SMALLER V8 DROPPED IN 1966, REPL. BY 307 CID (200 HP) IN 1968)

SEDAN PRICE $2542 ('60-61)
$2642 ('62-63) $2814 ('64)
$2793 ('65) $2874 ('66-67)
$3221 ('68) $3290 ('69)

6-CYL. LOW-PRICED "SUPERBA" SERIES AVAIL. 1959 TO 1963.

PRODUCTION PEAK OF 8173 IN 1962 (DOWN TO 5417 IN 1969)

CHEVELLE (NEW) by Chevrolet

64

INTERIOR (MALIBU)

(2-DR. WAGON ALSO AVAIL.)

194 or 230 cid 6 (120 or 155 HP @ 4400 RPM)

ALSO 283 cid V8 (195 or 220 HP @ 4800 RPM)

300

$2231. UP

PRODUCTION : MALIBU : 149,000
300 : 68,300 ; MALIBU SS : 78,800

MALIBU (ABOVE)

115" W.B.

$2376. (6)
($2484. (V8)

1965 PRODUCTION :
300 : 31,600 ; MALIBU : 152,200 ; MALIBU SS : 101,577
MALIBU SS-396 : 201

$2156. ~ 2858. PRICE RANGE

$2269.

300 DELUXE 13369 (6)

$2156.
300 2-DR. 13311 (6)

300 2-DR. 6-PASS. WAGON
(MALIBU 4-DR. WAGON has CHROME STRIP ALONG SIDE.)

MALIBU SS

new HORIZONTALLY-SPLIT GRILLE

65

MALIBU SS
13767 (6) $2750. CVT.
13867 (V8)   2858. "

194 or 230 cid 6 (120 or 140 HP @ 4400 RPM)
ALSO :
(283 cid V8 AVAIL. ONLY WITH 195 HP @ 4800 RPM)
3 new 327 cid V8s (250, 300, or 350 HP)

69

FULL COIL SUSPENSION

CHEVELLE

MALIBU CVT.= $3030.
(241,600 MALIBU MODELS BLT.)

MALIBU

H/T SPT. CPE.
$2821.

194 cid 6 (120 HP)
230 cid 6 (140 HP)
283 cid TURBO FIRE V8
(195 or 220 HP)
327 cid V8
(275 HP)
396 cid V8
(325 or 360 HP)

300 SERIES has
PLAINER REAR DECK w/o CHROME ORNAMENTATION.
6.95/7.35 × 14 TIRES

66 new GRILLE

(72,300 SS MODELS BLT.)
(SS CVT. ALSO AVAIL.)

H/T $2434. UP

MALIBU CVT.
2637. UP

MALIBU

(227,000 MALIBU MODELS BLT.)

TOTAL 1967 PRODUCTION:
375,831

67 new GRILLES and TAIL-LIGHTS

7.35 × 14 TIRES

$2825. **SS 396**

note "SS 396" IN GRILLE CENTER

SS 396

Turbo-Jet V8

H/T

(63,000 SS MODELS BLT.)

LARGER STD. 307 CID V8
(200 HP) USES REG. GAS
230 CID STD. 6
(140 HP)

CHEVELLE

MALIBU

1968 PRODUCTION:
432,302

68 (RESTYLED)

AT CENTER:
Chevelle Nomad Custom
WAGON (new)
$3303. (3-ST.)

new WB
112" 2 DR.
116" 4 DR.
(THROUGH '77)

WITH OPT. VINYL TOP

SS-396
H/T = $3249.

$2458. ~ 3266. PRICE RANGE

$2601. UP

FRONT VENT WINDOWS ELIMINATED

STD.
230 CID 6
(140 HP)
307 CID V8
(200 HP)

69 new GRILLES

new
LOCKING STEERING COLUMN
and TRANS. LEVER

MALIBU H/T
$3025.
$3372 WITH OPTIONAL ILLUS.

SS-396 PACKAGE
(also NOMAD,
GREENBRIER,
CONCOURS,
CONCOURS EST.
WAGONS AVAIL.)
116" WB

4530·01

CHEVROLET
(SINCE 1912)

A DIVISION OF GENERAL MOTORS CORP.

DK "FLEETMASTER" has CHROME TRIM AROUND WINDOW MOULDINGS.

SPORT SEDAN (73,746 BLT.) $1280.

6-CYL. (216½ CID ON ALL 1937 to 1949 CHEVROLETS and ON 1950 to 1952 w/o POWERGLIDE AUTO. TRANS.) (CARS)

90 H.P.

(CHEVROLET TRUCKS ALSO AVAILABLE)

TOTAL 1946 PROD.= 398,028

FASTBACK ONLY ON THIS "AERO" MODEL.

DK "FLEETLINE" AEROSEDAN (2-DR.) (57,932 BLT.)

$1249.

HORIZONTAL CHROME STRIPS ON FLEETLINE FENDERS

"46 MEDALLION

NEW POSTWAR GRILLE

DU "STYLEMASTER" NO CHROME TRIM AROUND WINDOW MOULDINGS.

SPORT SEDAN (75,349 BLT.) $1205.

PRICE RANGE: $1098. (BUSINESS COUPE, DU) TO $1712. (STATION WAGON, DK, w. WOODEN BODY)

CHEVROLET

FLEETLINE AERO

$1313.

FLEETMASTER SPORT SEDAN $1345.

FLEETLINE SPORTMASTER SED. $1371.

FLEETMASTER STATION WAGON (4-DOOR, 8-PASS.) $1893.

EJ, EK

47

new GRILLE has PROTRUDING CENTER SECTION

new MEDALLION

THIS DELCO STEERING-COLUMN RADIO CONTROL BOX IS RARE! MOST HAVE RADIO CONTROLS ON LOWER SECTION OF DASH.

FLEETMASTER CONVT. $1628.

	PRODUCTION
EJ STYLEMASTER	
4 DR. SPORT SEDAN	42,571
2 DR. TOWN SEDAN	88,534
SPORT (CLUB) COUPE	34,513
BUSINESS COUPE	27,403
EK FLEETMASTER	
4 DR. SPORT SEDAN	91,440
2 DR. TOWN SEDAN	80,128
SPORT (CLUB) COUPE	59,661
CONVERTIBLE	28,443
STATION WAGON (8-PASS.)	4912
EK FLEETLINE	
4 DR. SPORTMASTER SEDAN	54,531
2 DR. AERO SEDAN (FASTBACK)	159,407

$1439.

DASH

1948 CHEVROLET "FLEETMASTER" Four Door Sedan

(93,142 BLT.)

48

FJ, FK

↑ USUAL LOCATION OF RADIO

FJ STYLEMASTER 2-DR. TOWN SEDAN $1313.

(70,228 BLT.)

new "T"-SHAPED PIECE ADDED AT CENTER OF GRILLE

2144 FK FLEETLINE AERO (211,861 BLT.) $1434.

PRICE RANGE: $1244. TO $2013.
90 HP (1941 THROUGH 1949) @3300 RPM

OFFICIAL PACE CAR AT 1948 INDIANAPOLIS 500 RACE

CHEVROLET

STYLELINE

2152
2-DR.

FLEETLINE

new FASTBACK
4-DR. (2153)

2134
CVT.
$1857.
(32,392
BLT.)

← METAL-BODIED WAGON

PRICES
START AT
$1339.

(STYLELINE SPEC. BUS. CPE.)

GJ, GK

49

TOTALLY RESTYLED

GJ = SPECIAL
(1500 SERIES)

GK = DE LUXE
(2100 SER.)

1949 TRUNK LID has SMALL "T" HANDLE WHICH TURNS.

2102
2-DR. TOWN SEDAN

2103
4-DR.
SPORT
SEDAN

STYLELINE

2124
SPORT
CPE.

VERTICAL PIECES IN LOWER HALF OF GRILLE

new SHORTER 115" WB (THROUGH '57)

1949 HUBCAP has RED CENTER.

6.70 x 15

new PONTOON-STYLE REAR FENDERS

new DASH

all-new INTERIOR (LEFT AND RIGHT VIEWS)

DLX.
MODELS have CHROME AROUND WINDOWS and on FRONT FENDERS

CHROME (DLX.)
BLACK RUBBER (SPEC.)

DASH

2 VARIETIES OF STATION WAGON MODELS AVAILABLE (4 DR.)
#2109 (WOOD BODY) (3342 BLT.) ('49 ONLY)
or 2119 (STEEL ") (6006 BLT.) ('49 ON)
($2267. FOR EITHER ONE)

CHEVROLET

$1741. (76,662 BLT.)

new "Bel-Air" 2-DR. HARDTOP has WIDE BACKLIGHT → 2154

DASH

1950 TRUNK LID has new RE-DESIGNED HANDLE.

$1529. ↓

HJ, HK

50

TOTAL 1950 PRODUCTION: 1,498,590

new AUTOMATIC TRANSMISSION AVAILABLE

→

First low-priced car with **POWER***Glide* No-Shift driving *

FEWER VERTICAL PCS. IN 1950 GRILLE.

HJ = SPECIAL
HK = DE LUXE

2103 STYLELINE DLX. SPT. SEDAN (316,412 BLT.) (BEST-SELLING MODEL)

PRICE RANGE:
$1329. TO $1994.

* = POWERGLIDE SOMEWHAT LIKE BUICK'S "DYNAFLOW." (NOT INCLUDED IN ABOVE PRICES)

FLEETLINE DELUXE #2152 2-DR. SEDAN (189,509 BLT.)

BACK SEAT (4-DR.) STYLELINE DLX.

$1482. FOR EITHER →

1950 HUBCAP has YELLOW CENTER.

The Styleline De Luxe 2-Door Sedan (248,567 BLT.)

CHEVROLET

"See the U.S.A. in your Chevrolet..."

MODEL 2154

BEL AIR H/T (103,356 BLT.)

$1914.

1500 SERIES SPEC. STYLELINE PRICES START AT $1460.

STYLELN. DLX. 2102 2-DR. (262,933 BLT.)

$1629.

new GRILLE and SIDE TRIM

JJ, JK

51

FLEETLINE DLX. 2152 2-DR. (131,910 BLT.)

$1629.

DE LUXE

Chevrolet

MODEL 2103 4-DR. STYLELINE DLX. (380,270 BLT.)

$1680.

TOTAL 1951 PROD.: 1,229,986

NEW Safety-Sight Instrument Panel

INTERIOR

NEW Modern-Mode Interiors

76

CHEVROLET

$1620.

$1707.

#1524 STYLELINE SPECIAL SPORT CPE. IS SCARCE, WITH ONLY 8906 BLT. →

(37,164 BLT.)

STYLELINE SPECIAL (has MINIMUM of CHROME TRIM)

2-DR. ONLY

FLEETLINE DLX. (NO MORE FLEETLN. SPECIAL OR 4-DRS.)

NEW

26 Exterior Colors and two-tone color combinations to choose from.

New Softer, Smoother Ride with new and improved shock absorber action.

Improved Carburetion with Automatic Choke in Powerglide models.

New Centerpoise Power is smoother — "screens out" engine vibration.

Color-Matched Two-Tone Interiors bring new beauty to De Luxe models.

52

KJ, KK
SPECIAL DELUXE

STYLELINE DE LUXE 2-DR.

$1707.

(215,417 BLT.)

new 5 RIDGES RUN DOWN CENTER HORIZ. MEMBER of GRILLE.

$1726

new MEDALLION

$1519. TO $2281.
PRICE RANGE

(2 VIEWS)

STYLELINE DE LUXE SPORT COUPE (ABOVE)
(36,954 BLT.)

$1992.

BEL AIR (IN STYLELINE DLX. SERIES) H/T (74,634 BLT.)

TOTAL 1952 PRODUCTION:
818,142

1952 IS FINAL YEAR FOR STYLELINE and FLEETLINE MODEL NAMES.

CHEVROLET

CLUB CPE. $1620.

BUS. CPE. $1524.

150

150 HAS NO SIDE CHROME

210 HAS SINGLE SIDE CHROME STRIP

210 2-DR.

$1707.

$1761.

$2010.

ABOVE: 210 SEDAN (IN SAN FRANCISCO, CALIF.)

$1874.

53

(TOTALLY RESTYLED)

FIRST 1-PC. WINDSHIELD ON A CHEVROLET SINCE 1935!

BEL AIR IS NEW TOP-OF-LINE SERIES.

Handyman (two of them) 6-PASS.

150

station wagons

210

BEL AIR SEDAN (INTERIOR)

H/T

$2051.

$2123.

235 CID ENGINE (THROUGH '62, ON 6-CYL.) 108 OR 115 HP @ 3600 RPM)

BEL AIR (note EXTRA TRIM and CONTRASTING COLOR STRIP on REAR FENDER.)

WITH IMITATION WOODGRAIN TRIM

DASH

$2175.

$2273.

Townsman 8-PASS.

150 SPECIAL	210 DELUXE	
PRODUCTION		2134 CONVERTIBLE (5617)
1502 2.DR. SED. (79,416)	2102 2.DR. SED. (247,455)	2154 H/T CPE. (14,045)
1503 4 DR. " (54,207)	2103 4 DR. " (332,497)	**240 BEL AIR** = 2402 2.DR. SED. (144,401)
1504 BUSINESS CPE. (13,555)	2109 HANDYMAN WAGON (18,258)	2403 4 DR. " (247,284)
1509 WAGON (22,408)	2119 TOWNSMAN 8-PASS. " (7988)	2434 CONVERTIBLE (24,047)
1524 CLUB COUPE (6993)	2124 CLUB CPE. (23,961)	2454 H/T CPE. (99,028)

CHEVROLET

new MEDALLION
new TAIL-LIGHTS

54

5 VERTICAL GRILLE PCS. INSTEAD OF 3

Push Button Window Controls†
Toe-Touch Power Brake Pedal†
Extra-Easy Power Steering†
Powerglide Automatic Transmission†
Push Button Door Latches†
Push Button Door Locks (Keyless Locking)
Push Lever Heater Controls†
Push Button Radio Controls†
Automatic Dome Light Switches†
Push Button Glove Compartment Lock (Automatic Light†)
Pull Knob Light Switch
Finger-Touch Horn Blowing Ring†
Pull Knob Ventilation Controls
Key-Turn Starter (Automatic Choke)
Turn Knob Wind-shield Wiper Control (Push Button Washer†)
Toe-Touch Accelerator Treadle
Push Button Headlight Dimmer
Lever Action Direction Signal Control (Automatic Return†)
Push Button Automatic Seat Adjustment Control†

Advanced Chevrolet Engineering brings
CYBERNETIC CHEVROLET
(Cybernetic = Automatic Control)

$1782.

210 DELRAY COUPE 2124
(66, 403 BLT.)

$1884.

2403
BEL AIR SEDAN
(248, 750 BLT.)

BEL AIR

new OBLONG PARKING LIGHTS

115 HP @ 3700 RPM
OR 125 HP @ 4000 RPM

DASH

PRICE RANGE = $1539. TO $2283.
"150" 2-DR., 3-PASS.   "240" BEL AIR
UTILITY SEDAN   TOWNSMAN 8-PASS.
(10,770 BLT.)   (8156 BLT.)

79

CHEVROLET

185,562
BEL AIR H/Ts
BLT.

150 UTILITY SED.
(3-PASS.)
(11,196 BLT.)
$1593.

#1512

"ONE-FIFTY" HANDYMAN
$2030.
(17,936 BLT.)
#1529

BEL AIR
$2206.

#2103
$1819.
(317,724 BLT.)
THE "TWO-TEN" 4-DOOR SEDAN in Skyline Blue.

(ABOVE)
210
HANDYMAN
(28,918 BLT.)
$2079.

(TOTALLY RESTYLED)

55

$2127.
(82,303 BLT.)
#2109
210 "TOWNSMAN" WAGON

new V-8

(1ST CHEVROLET V8 SINCE 1918!)

(ABOVE) BEL AIR CVT.
(41,292 BLT.)

A = GAS GAUGE   C = WATER TEMP.

WITH CONVENTIONAL TRANSMISSION

265 CID V8 162 HP @ 4400 RPM, 170 HP, OR 180 HP @ 4600 RPM

235½ CID 6
123 HP @ 3800 RPM OR 136 HP @ 4200 RPM

DASH

REVERSE SECOND
NEUTRAL
FIRST (LOW) THIRD (HIGH)
CLUTCH PEDAL

2409
$2262.

THE BEL AIR BEAUVILLE
(24,313 BLT.)

new "NOMAD" 2-DR. WAGON

(CHROME STRIPS RUNNING DOWN TAILGATE.)

(8386 BLT.)

CVT. IS PACE CAR AT 1955 INDY 500 RACE

$2472. (6)

CHEVROLET

1529
THE "ONE-FIFTY" HANDYMAN
2 doors, 6 passengers, versatile and thrifty.

(13, 487 BLT.)
$2171.

56

2129
THE "TWO-TEN" HANDYMAN
2 doors, 6 passengers, all-vinyl interior.

2119
THE "TWO-TEN" BEAUVILLE
4 doors, 9 passengers.

2109
THE "TWO-TEN" TOWNSMAN
4 doors, 6 passengers, loads of cargo space.

BEL AIR 4-DOOR HARDTOP and interior

AIR COND. DETAIL

(103,602 BLT.)

MODEL 2413

$2329.

Now in the low price field...

All components are located "up front" ... out of sight and out of the way! Harrison air conditioning is available on four great GM cars—Chevrolet, Pontiac, Oldsmobile and Buick.

AIR CONDITIONING!

2419
BEL AIR BEAUVILLE 9-PASS. WAGON (13,279 BLT.)
$2482.

new SMALL ROUND LENSES IN TAIL-LIGHTS

$1912.

210

AA-1956

2102 2-DR. S1D.
(205,545 BLT.)

2403
$2068.
BEL AIR SEDAN
(269,798 BLT.)

CORVETTE

2429
NOMAD

new FULL-WIDTH GRILLE

2402
BEL AIR 2-DR.

(104,849 BLT.)
$2025.

235½ CID 6 has new 140 HP

265 CID V8 AVAIL. WITH 162, 170, 205 OR 225 HP

6.70 x 15 TIRES

CHEVROLET

PRICES START AT $1885.
(150 2-DR. UTIL. SED.)

150

$2048.

#1503 SEDAN   (52,266 BLT.)

210

#2403 4-DR. SED. (254,331 BLT.)

$2290.

BEL AIR

(27,375 BLT.)

$2580.

1957 IS 3RD AND FINAL YEAR THAT THE NOMAD IS A SUPER-DELUXE 2-DOOR SPORT WAGON.

$2757. (6)

NEW TRIPLE-TURBINE TURBOGLIDE*
It's the last word in automatic drives. Super-smooth— and there's even a HILL-RETARDER position on the selector, for safer control on the steepest down grades!

57

#2429   1957   (6103 BLT.)

NOMAD and BEL AIR have new ANODIZED REAR FENDER PANEL.

4-DOOR WAGON (#2409 "TOWNSMAN")

new GRILLE COMBINED with BUMPER

2454
$2299.   2-DR. H/T   (166,426 BLT.)

DASH

HEADLIGHT-HOOD AIR VENTS

COMMAND POST CONTROL PANEL

new 7.50 × 14 TIRES

ENGINES = 235 1/2 CID 6 has 140 HP.
265 CID V8 has 162 HP.
new 283 CID V8 ALSO, AVAILABLE WITH 185, 245, 250, 270 or 283 HP.

4-DR. H/T SPORT SEDAN

$2464.   2413   (137,672 BLT.)

1-USA

new TAILFINS and ARCHED TAIL LIGHTS

CHEVROLET

NOMAD 6-PASS. 4-DR.

DASH

BEL AIR

4-DR., 6 or 9-PASS. BROOKWOOD

$2571. and up

new 117½" WB (1958 ONLY)

TOTALLY RESTYLED

58

BISCAYNE

235 CID 6 has 145 HP @ 4200 RPM

2-DR. 6-PASS. YEOMAN

#1191 (6 CYL.) YEOMAN IS LOWEST-PRICED CHEV. WAGON, AT $2413.

CROSS-SECTION OF "TURBO THRUST" V8 ENGINE

283 OR 348 CID V8s (TO '62) (185 TO 280 HP)

new MODEL SERIES IN 1958
DELRAY = $2013. up
BISCAYNE = 2236. up
BEL AIR and
IMPALA = $2386. to 2841.

IMPALA (new)

#1867 CVT. $2841.

IMPALA DASH

#1847 H/T $2693.

new WAGON TAILGATE

IMPALAS have 6 REAR LIGHTS, AND EXTRA "AIR SCOOP" DECORATIONS.

IMPALA

CHEVROLET

1—*Biscayne Utility Sedan.* Chevy's prices start right here—a handy, handsome 2-door with 31 cu. ft. of cargo space behind front seat.

2—*Brookwood 2-Door,* Chevrolet's lowest priced wagon, is as dutiful as it is beautiful. Seats 6, holds up to 92 cu. ft. of cargo.

3—*Impala 4-Door,* most elegant family sedan in the line, makes you wonder why anyone would want a car that costs more.

4—*El Camino* combines stunning passenger car styling with the load space of a pickup. Good looks never carried so much weight!

5—*Impala Convertible.* Chevy's got a special formula for carefree top-down fun.

6—*Biscayne 2-Door.* This beauty's the lowest priced 6-passenger Chevy you can buy!

7—*Nomad 4-Door,* 6-passenger station wagon—finest of Chevrolet's 5 wonderful wagons.

8—*Bel Air 4-Door.* As luxurious as it looks, yet priced just above Chevy's thriftiest sedans.

9—*Brookwood 4-Door.* Chevy's lowest priced 4-door wagon seats 6, holds 92 cu. ft. of cargo with rear seat down.

10—*Bel Air 2-Door,* distinctively styled inside and out, carries a price tag just a notch above Chevy's thriftiest 2-door sedan.

11—*Impala Sport Sedan.* Here's a 4-door hardtop with the kind of looks and luxury you'd expect only on the most expensive makes.

12—*Kingswood 4-Door,* 9-passenger station wagon, offers rear-facing third seat and power-operated rear window at no extra cost.

13—*Impala Sport Coupe.* It's one of Chevy's full series of elegant Impalas for '59. And you won't find a handsomer hardtop anywhere!

14—*Parkwood 4-Door,* 6-passenger station wagon, distinctively trimmed inside and out, priced a shade above the thrifty Brookwoods.

15—*Bel Air Sport Sedan.* It's Chevy's lowest priced hardtop—and it makes beautiful sense!

16—*Corvette.* Take the wheel of America's only authentic sports car and treat yourself to the snappiest, happiest driving you've known.

17—*Biscayne 4-Door,* thriftiest 4-door sedan in the line, is another big reason

BROOKWOOD

135 TO 315 HP

PRICE RANGE
$2160.
TO
$3009.

BEL AIR

$2440.
and up

$2891.
and up

NOMAD
4-DR., 6-PASS.

HUGE *new* TAIL-LIGHTS

59
(TOTALLY RESTYLED)

BIG "GULL WING" REAR DECK

1959

GENERATOR AND OIL PRESSURE INDICATOR LIGHTS • TEMPERATURE GAUGE • SPEEDOMETER • BRIGHT BEAM INDICATOR • FUEL GAUGE • CLOCK

BRAKE SIGNAL LIGHT • LIGHT SWITCH • WINDSHIELD WIPER SWITCH • RIGHT AND LEFT TURN SIGNALS • ODOMETER • HEATER CONTROL • IGNITION SWITCH

new INSTRUMENT PANEL

1737 (6)
1837 (V8)

IMPALA SPORT COUPE (H/T)
$2599. (6) **$2717.** (V8)

new 119" WHEELBASE (THROUGH 1970)

84

CHEVROLET

BISCAYNE
(NO SIDE CHROME)

NOMAD

60

KINGSWOOD

PRICE RANGE: **$2230.** TO **$2996.**

BEL AIR

BEL AIR

#1511
2. DR.
SEDAN
$2384.

new GRILLE

#1737
IMPALA
SPORT CPE.

$2597.

Impala 4-Door Sport Sedan

"GULL-WING"
REAR STYLING
MODIFIED.
new ROUND
TAIL LIGHTS and
BACKUP LIGHTS.

DASH

#1739 SPORT SEDAN
4. DR. H/T **$2662.** UP

ENGINES = 235½ CID 6 (135 HP)
283 CID V8 (170 HP)
348 CID V8 (250 OR 335 HP)

CHEVROLET

$2316. (6)

BROOKWOOD
$2653. up

$2230. up

BISCAYNE

NOMAD

PRODUCTION:
1,201,811
135 to 360 HP
new 409 cid V8
JOINS OTHER
ENGINES

$3099. 9-PASS.
V8

new ROOFLINE
(SPT. CPE.)

BISCAYNE
(PHOTO) $2423. (V8)

(ARTIST'S
CONCEPTION, TOP LEFT)

(RESTYLED)

61

(64,600
BLT.)
CVT.

$2954. (V8)

(HT)
SPT.
CPE.

BEL AIR

$2489. up

new
ROOFLINE →

BEL AIR SPT. SED.

$2230.
TO $3099.
PRICE RANGE

IMPALA
$2590. (6)

$2554. up

IMPALA
H/T

$2597.
(6)

INSTRUMENT PANEL

LIGHT CONTROL
SWITCH

CIGARETTE LIGHTER
AND ASH TRAY

RADIO CONTROLS

LEFT VENT
CONTROL

WIPER AND WASHER
CONTROL

HEATER
CONTROLS

RIGHT VENT
CONTROL

IGNITION
SWITCH

GLOVE BOX
AND LOCK

$2725. UP

BISCAYNE $2378. →

FINAL 235 CID 6 has 135 HP @ 4000 RPM

$2819. UP

BEL AIR SPT. CPE. ROOFLINE

DASH

$2510. UP

BEL AIR

$3026.

1962

IMPALA

new GRILLE

M-1042

AG-1400

IMPALA has ALUMINIZED PANELING AROUND TAIL-LIGHTS.

OUTER-EDGE TAIL-LTS. DO NOT OPEN WITH TRUNK.

IMPALA

JET-SMOOTH RIDE

(IMPALA I.D.)

IMPALA 4 DR. H/T

62 RESTYLED

ENGINES
230 CID
6
(140 HP)
283 CID V8 (170 HP)
new 327 CID V8 (250 OR 300 HP)
409 CID V8 (380 OR 409 HP)

$ **2734.** (6)
2841. (V8)

PRODUCTION:
1,495,476

DASH (CLOSER VIEW)

CONTROLS

WIPER WASHER — IGNITION — VENT — HEATER — LIGHTER — RADIO — ASH TRAY — GLOVE BOX

1963 PROD.: 1,625,931

$2376. (6)   BISCAYNE

DASH

$2519.

IMPALA SPORT SEDAN

BEL AIR

PRICE RANGE: **63**
$2558. TO $3417.

$2732. (6)
$2839. (WITH V8)

note CONVERTIBLE-STYLE "CREASES" STAMPED INTO STEEL ROOF OF THIS IMPALA SPORT COUPE

new DIP IN MIDDLE OF DECK LID ON 1963 MODELS
$2786.

new
230 CID 6
(140 HP @ 4400 RPM)
283 CID V8 (170 HP)
327 CID V8
(250 OR 300 HP)
409 CID V8
(340 OR 425 HP)
@ 6000

#1847 H/T SPT. CPE. (V8)
$2774.

88

1964 JET-SMOOTH CHEVROLET

BISCAYNE

417.
UP

BISCAYNE

BISCAYNE
2-DR.

$2363.

$2590.
(WEST)

new STRAIGHT-ACROSS
DECK LID with CENTER RIDGE

BEL AIR

$2519.
UP

64

PRICE RANGE = $2363. TO $3196.

RESTYLED

BEL AIR 4-DR. SED.
#1569 (6)
1669 (VB)

283, 327, 409 cu V8
ENGINES, SAME
SIZES AS IN
'63   6 = 140 HP
V8s = 195, 250, 300,
340, 400 OR
425 HP

3.08 TO
4.56
GEAR RATIOS

REAR
DECK
DETAILS
(IMPALA)

1964
PRODUCTION: 1,420,304

THERE'S 5 IN
64 CHEVROLET
CHEVROLET · CHEVELLE · CHEVY II · CORVAIR · CORVETTE

C-1964

SPT. SEDAN
4-DR. H/T

9-PASS.
WAGON
1745
CYL.

$2742. UP

IMPALA

$3073.

H/T

IMPALA SS

1447
new
IMPALA
SS

3185.,
WEST COAST)

$2947.
ELSEWHERE

IMPALA V8
CONVERT.
$3035.
(SS,
$3196.)

C-1964

new GRILLE

DASH

89

CHEVROLET

$2669.. 4 DOOR BISCAYNE

$2519. UP

7.35 x 14 TIRES

(107,700 BISCAYNE 6s; 37,600 BISCAYNE V8s)

BEL AIR

(163,000 BEL AIR V8s; 107,800 6s)

$2742. (6)

IMPALA 4 DR. H/T

$2850. (V8)

IMPALA 3-SEAT WAGON

FY-1588

$3181.

8.25 x 14 TIRES ON WAGONS

1965

803,400 IMPALAS BLT. (6 or V8)

POPULARLY REFERRED TO AS THE "COKE BOTTLE" PROFILE

$2947.

#16637 IMPALA SS H/T

(243,100 IMPALA SS H/Ts and CVTS., 6 or V8)

PRICE RANGE = **$2363.** TO **$3212.**

65 (TOTALLY RESTYLED)

1965 PRODUCTION: 1,821,266

CONTROLS

VENT — LIGHTS — WIPER WASHER — IGNITION SWITCH — LIGHTER — ASH TRAY — RADIO — HEATER — GLOVE BOX — VENT

243,100 IMPALA SS BLT. (H/T or CVT., 6 or V8)

IMPALA SUPER SPORT

(WITH OPTIONAL EQUIPMENT)

AVAIL. with VINYL TOP COVERING

with SPORT WHEEL COVERS →

WEST COAST: **$3210..**

Chevrolet

CHEVROLET

Caprice DASH

REAR VIEW

$3800. (3-SEAT)

$3347. Caprice Custom Wagon

(1966 MODELS INTRO. THURSDAY 10-7-65)

IMPALA HUB CAP

66

IMPALA

new GRILLE, BUMPER and TAIL-LIGHT

NEW

119" WB
7.75/8.25 x 14 TIRES

STD. ENGINES
250 CID 6 (155 HP)
283 CID V8 (195 HP)

CAPRICE NOW A SERIES, INSTEAD OF A PACKAGE AS IN 1965.

MODEL SERIES: BISCAYNE (122,400 BLT.) $2379.~2877.;
BEL AIR (236,600 BLT.) $2479.~3053.; IMPALA (654,900 BLT.)
$2678.~3189.; IMPALA SS (119,300 BLT.) $2842.~3199.;
CAPRICE (181,000 BLT.) $3000.~3347.; WAGONS (185,500 BLT.)

* CAPRICE
V8 Custom Series

note HORIZONTAL BANDS ACROSS CAPRICE TAILLIGHTS (UNLIKE TAILLIGHTS OF BISCAYNE, BEL AIR OR IMPALA MODELS)

"CUSTOM SEDAN"
$3516.

"4 DR. H/T

$3063.

*-(CAPRICE INTRO. 1965, AS A $242 OPT. PKG.)

$3000.

$3453.

CAPR. CUST. CPE. ROOFLINE (New)

DENOTES A 327 CID V8 (250 OR 300 HP)

WEST COAST PRICES SHOWN IN SMALL PRINT.

CAPRICE SERIES ALSO INCLUDES 2 WAGONS:
2-SEAT = $3234.   3-SEAT = $3347.

1967 CHEVROLET PRICE RANGE : $2442. ~ 3413.

'67 Chevrolet gives you that sure feeling

$2971. UP

WEST COAST, $3469. UP

Biscayne 2484.
$3036. (WEST)

NO SIDE CHROME ON BISCAYNE

(INTRO. 9-29 66)

Impala

STANDARD ENGINES
155-hp Turbo-Thrift 250 Six
195-hp Turbo-Fire 283 V8
EXTRA-COST OPTIONAL ENGINES
275-hp Turbo-Fire 327 V8
325-hp Turbo-Jet 396 V8
385-hp Turbo-Jet 427 V8

'67 IMPALA

Impala SS
$3350. (WEST COAST)

67 new GRILLE

$3192. (WEST)

Caprice Custom Sedan

$3477. (WEST COAST)

1967

$2845. (V8)

Impala Sport Coupe

$3130.   TOTAL 1967 PRODUCTION : 1,201,700 (CALENDAR YEAR : 1,150,264)

LARGEST ENGINE (1967 and 1968) IS 427 CID V8 (385 HP) (OPT.)

TOTAL 1968 PRODUCTION : 1,235,800 (CALENDAR YEAR : 1,217,255)

new GRILLE IS HORIZONTALLY BISECTED BY BUMPER CROSS BAR

68

(WEST COAST) $3809. UP

CAPRICE

CUSTOM SEDAN $3621.

new ROUND TAIL LTS. IN BUMPER

CUSTOM CPE.
formal $3021.

STD. 250 CID 6 (155 HP) 307 CID V8 (200 HP)
(327, 396, 427 CID V8s AVAIL.)

$3371. (WEST)

Impala coupes

$2968.

1968

note SMALL new SIDE LIGHTS

$2581. ~ 3570. PRICE RANGE (INTRO. 9-21-67)

Fastback SPORT CPE.

BE SMART! BE SURE! BUY NOW AT YOUR CHEVROLET DEALER'S.

CHEVROLET

ALL ENGINES
EXCEPT 427 CID V8s
USE REGULAR GAS.

$2981. UP
$3427.
(WEST)

GM

new PLASTIC GRILLE

Impala
H/T $2927.
119" WB

69

(INTRO. 9-26-68)
(RESTYLED)

new KINGSWOOD ESTATE (3-SEAT) $4019.

Caprice.

ENGINES:
250 CID 6 (155 HP); 327 CID V8 (235 HP);
350 CID V8 (255 or 300 HP);
396 CID V8 (265 HP);
427 CID V8 (335 or 390 HP)

$2645.~3678.
PRICE RANGE

"walk-in wagon"
has 2-WAY TAILGATE

wagon

327 CID V8 AVAIL.

new CONCEALED HEADLIGHTS

OPTIONAL "LIQUID TIRE CHAIN" TRACTION IMPROVER SQUIRTS ONTO REAR WHEELS AT THE PUSH OF A BUTTON.

TOTAL 1969 PRODUCTION: 1,227,600

new BROOKWOOD, TOWNSMAN, KINGSWOOD and KINGSWOOD ESTATE WAGONS

note THE FLARED FENDERS
2 DR. H/T

Caprice $3294.

Putting you first, keeps us first.

(4-DR. H/T, WAGON ALSO IN CAPRICE SERIES)

NO SIDE CHROME ON 100

Chevy II (new)

$2122.

$2041.

REAR DETAILS

COMPACT CAR **by Chevrolet**

(STARTS 1962)

300 $2084.

$2475.

Nova

WAGON

Wagon

6.50 × 13 TIRES ON WAG., 6.00 × 13 ON OTHERS

CONVERTIBLE (SHOWING DASH, INTERIOR DETAIL)

0435 NOVA WAGON 400 (6 CYL.) $2497.   AUTO. TRANS. = $167.

0467 NOVA 400 CVT.   (23,741 BLT.)

110" W.B. (THROUGH '67)

ENGINES =
153 CID 4 (90 HP) ($62. LESS)
194 CID 6 (120 HP)
(2 CHOICES THROUGH 1963)

0437 (6 CYL.)

NOVA 400 H/T

$2254.

(59,586 BLT.)

POWER STEERING AVAIL.

Nova

PRICE RANGE: **62** $2051. TO $2793. (WEST COAST)

REAR and FRONT FENDER and WHEEL COVER DETAIL (NOVA 400)

DASH

(AS SEEN FROM REAR OF WAGON)

CHEVY II

100
$2003.
$23/3.
(WEST COAST)

0311 SEDAN
$2084.

300

$2395.
(WEST COAST)

(FINAL YEAR
FOR 300
SERIES)

$2494.
$2710.
(WEST COAST)

63

0435
WAGON

NOVA
400

0449
SEDAN
$2235.

NOVA CVT.
400   0467

($2687. WEST COAST)

new GRILLE ←

AIR
COND.
$317.
EXTRA

$2472.
(SUPER
SPORT PKG.
w. BUCKET
SEATS,
$161.
EXTRA)

120 HP
(6 CYL.)
SINCE '62)

NEW V8 POWER

with OPTIONAL
283 C.I.D. V8
(195 HP) ALSO
USED IN CHEVELLE
and CHEVROLET)

4 and 6 CYL.
CONTINUE

64

| LIGHT SWITCH | WIPER SWITCH AND WASHER BUTTON | IGNITION SWITCH | RADIO CONTROLS | GLOVE BOX LOCK | GLOVE BOX |

MANUAL CHOKE
(4 CYLINDER ENGINE)

AIR VENT
CONTROL

PARKING
BRAKE

CIGARETTE
LIGHTER

ASH
TRAY

AIR   TEMP   DEF
HEATER CONTROLS

AIR VENT
CONTROL

$2503.

435
NOVA
400
WAGON

0110 4 CYL. 2 DR. 100 $2011.

BU·8715

MORE '64s ON NEXT PAGE

CHEVY II

INTERIOR (H/T)

64 (CONT'D.)

SPT. CPE.

0447
NOVA SS
(has THIS EMBLEM)

$2433.

NOVA

0469
SEDAN

$2243.

AVAILABLE MODELS = 100 2-DR. (4) $2011.; (6) $2077.; 100 4-DR. (4) $2048.; (6) $2115.
100 4-DR. WAGON (6) $2413.
NOVA 400 (6) 4-DR. $2243.
4-DR. WAGON $2510.
H/T COUPE $2270. (NOVA SS $2433.)

65

new GRILLE,
new PLAIN RED TAILLIGHTS, SEP. BACKUP LIGHTS

100   11335
(TAILGATE OPEN)

100
11169 (4)
11369 (6)

Super Sport

11737
H/T

NOVA

(TAILGATE CLOSED)

11535 WAGON

11569
SEDAN

6.50 × 13 TIRES
7.00 × 13, WAGONS
6.95 × 14, S.S.

(1962 - 1979) **CHEVY II** by CHEVROLET   DELUXE MODELS KNOWN AS **Nova**

153 CID 4 (93 HP)
194 CID 6 (120 HP)
283 CID V8 (195 HP)
6.50 × 13 TIRES
(WAG. = 6.95 × 14)

Station Wagon

110" WB
(1962 - 1967)

66 new GRILLE; new TALLER TAIL-LTS.

NOVA H/ SUPER SPORT H/T
$2652. (WEST COAST)

2430. (6)   2535. (V8)
(6700 BLT.)   (16,800 BLT.)

1966 PRODUCTION:
155,726 (CALENDAR YR.)

LOWEST-PRICED "100" 2-DR. IS
$2028. ($2250.)
(WEST COAST)

$2090. ~ 2671. PRICE RANGE

1967 PRODUCTION:
135,884 (CALENDAR YR.)

STATION WAGON AVAIL. THROUGH '67, FROM $2478.

4-CYL = 90 HP
6-CYL = 140 HP
V8 = 195 HP

67

1967 GRILLE DIFFERENCE ILLUSTR. AT LOWER LEFT

NOVA SS H/T
$2487. UP

'67 Chevy II
The stylish economy car

NOVA 4 DR.
$2298. UP

(WAGONS NO LONGER AVAIL.

SS OPTION: $210. EXTRA

STD. 4-DR. (w/o SIDE CHROME)

$2261. UP

SS

68 (TOTALLY RESTYLED) new 111" WB (THROUGH '79)

1968 EASILY IDENTIFIED BY "CHEVY II" NAME ON TOP BORDER OF GRILLE.

ENGINES:
153 CID 4 (90 HP)
230 CID 6 (140 HP)
307 CID V8 (200 HP)
UP TO 295 HP AVAIL.

OPTIONAL BUMPER GUARDS

CHEVY Nova

153 CID 4 (90 HP); 230 CID 6 (140 HP)
307 CID V8 (200 HP)   SS OPTION: 350 CID V8 (300 HP)

(NOVA REPLACES CHEVY II NAME)

$2237. UP

69

7.35 x 14" TIRES

new COWL LOUVRES

"SS" GRILLE

CHEVROLET EMBLEM NOW APPEARS ON TOP BORDER OF GRILLE.

CHRYSLER

EMBLEM
Chrysler

ROYAL 6 (C-38-S)
WINDSOR 6,
TOWN and COUNTRY 6
SEDAN (C-38-W)
SARATOGA 8 (C-39-K)
NEW YORKER 8,
TOWN and COUNTRY 8
CVT. * (C-39-N)
CROWN IMPERIAL 8
(C-40)

*=100 T+C 8
SEDANS
ALSO

WHITE "BEAUTY RING"

(ABOVE = SARATOGA 8
SEDAN
$1863.

WINDSOR 6
CVT. $1861.

6 CYL. has 250.6 CID
(1942 THROUGH 1951)
114 HP @ 3600 RPM
(THROUGH '49)

STRAIGHT-8 has
323.5 CID (1935
THROUGH 1950)
135 HP @ 3400 RPM
(THROUGH '49)

WHITE
"BEAUTY RINGS"
USED, BECAUSE OF SCARCITY OF
WHITE SIDEWALL TIRES.

NEW YORKER 8 CLUB COUPE
$1948.

(3-WINDOW
BUSINESS CPES.
HAVE LONG
REAR DECKS!)

PRICE RANGE:
$1415. TO $4767.

46-48

CONTINUES NEARLY UNCHANGED
TO FEB., 1949

121½" WB (6)
127½" " (8)
145½" " (CROWN
IMPERIAL 8, THROUGH
'54)
139½" WB (6-CYL.
LIMOUSINE)

TEMP. BEAM TURN SIGNAL LITE OIL CLOCK
Chrysler
AMPS
C-38 SERIES
FUEL
SPEEDO.
HORN RING
SHIFT LEVER
DASH

BEAUTIFUL DASH
OF METAL, CHROME
and PLASTIC,
COLOR KEYED TO
BODY COLOR.

AT NIGHT,
SPEEDOMETER
NUMERALS
CHANGE COLOR
FROM GREEN
TO AMBER TO
ORANGE TO
RED AS THE
SPEED
INCREASES!

new
CAST-METAL
"HARMONICA
STYLE"
GRILLE

ROYAL 6 CLUB COUPE
$1551.

1946 CHRYSLER CORP.
CARS HAVE FLAT
DOOR LOCK COVERS;
1947-48 PROTRUDE.

"TOWN and COUNTRY"
MODELS,
NEXT PAGE

99

CHRYSLER

TOWN + CNTRY.
H/T (ONLY 7
BUILT)

CONVENTIONAL
CONVERTIBLE
INTERIOR →

46-48
(CONT'D.)

TOWN -
and -
COUNTRY

REAR DETAILS
(SEDAN)

(FIRST
1941-1942
T + Cs ARE
FASTBACK
HATCHBACKS.)

SEDANS =
MOST 1946-48 T+C
6 CYL., BUT SOME 8s ALSO.

TOWN and COUNTRY

$3123.
('48)

CONVT. (8-CYL.)

ONE-OFF
2-DR. BROUGHAM
(EXPERIMENTAL)

Natural wood

'46 - EARLY 1947
"TOWN and
COUNTRY"
MODELS have
GENUINE
WOODEN
PANELS of
ASH and
MAHOGANY.
(DK. PANELS
ON LATER
MODELS
ARE
DECALS)

CHRYSLER

WINDSOR 6

(55,879 BLT.)

(4524 BLT.)

CLUB COUPE

NEW YORKER 8

(C-45) 125½" WB (THROUGH '54)

(LENGTH EXAGGERATED)

$3206.

(1137 BLT.)

WINDSOR 6 LIMOUSINE has 139½" WB (73 BLT.) $3144.

NEW YORKER 8

ACTUAL LENGTH

(18,799 BLT.)

NEW YORKER 8 (C-46) 131½" WB (THROUGH '52)

ROYAL 6 and SARATOGA 8 ALSO AVAIL.

49

(TOTALLY RESTYLED FEB., 1949)

PRESTOMATIC FLUID DRIVE* TRANSMISSION
*gyrol Fluid Drive

CHRYSLER

PRICE RANGE: $2114. TO $5334.

$3970.

TOWN and COUNTRY 8 CONVERTIBLE (SAME SPECS. AS NEW YORKER) (1000 BLT.)

CROWN IMPERIAL 8 (C-47)

LIMOUSINE

(45 BLT.)

$5334.

145½" W.B.

(FINAL 1948 CHRYSLERS SOLD AS "EARLY 1949" DURING JAN. and FEB., 1949.)

CHRYSLER

(FINAL YEAR FOR CHRYSLER STRAIGHT-8.)

ROYAL 6, WINDSOR 6
(C-48)

SARATOGA 8,
NEW YORKER 8,
TOWN and COUNTRY 8
(C-49)

50

PRICE RANGE

$2114. TO $5334.

TOWN and CNT. 8
AVAIL. ONLY AS
H/T.

new GRILLE

NY 8

$4003.

$5334. *Crown Imperial*

LIMOUSINE

(C-50)

CRN. IMP. REAR COMP. has QUARTER WINDOWS.

8-CYL.
MODELS NOW
DEVELOP
135 HP @
3200 RPM.

$3674. ('51)
3839. ('52)

131½" W.B.

SARATOGA V8 (C-55)
125½" WB

NY V8
(C-52)
131½" WB

IMPERIAL V8
(C-54)

(CRN. IMP. IS C-53)

NEWPORT
H/T →

WINDSOR 6
(C-51-1)

FORMER STR.-8
REPLACED '51 BY
new V8 ENGINE (331.1 CID)
180 HP, O.H.V.)
@ 4000 RPM

new
WIN. DLX. 6
IS C-51-2

125½" W.B.
(LENGTH EXAGGERATED IN
THIS ADVERTISING
ILLUSTRATION)

new GRILLES,
WIDER REAR
WINDOWS IN
1951.

51-52

(new
"HYDRA-GUIDE"
POWER STEERING
AVAILABLE,
1951 ON)

6-CYL. ENGINE CHANGES
(1951) 250.6 CID, 116 HP
@ 3600 RPM
(1952) 264.5 CID, 119 HP
@ 3600 RPM

(CVT.) IS PACE CAR AT 1951 INDY 500 RACE

(MORE IMPERIALS, CUSTOM MODELS
NEXT PAGE)

K-310 CUSTOM-BUILT COUPE

CUSTM BLT.

125 ½" W.B. (ITALIAN GHIA BODY)

PHAETON   147½" W.B. (FOR PARADE USE, ETC.)

Imperial

BY CHRYSLER

$4042. ('51)
4224. ('52)

A VARIETY OF ROSE WAS NAMED "CHRYSLER IMPERIAL."

IMPERIAL V8 (C-54)

NEWPORT IMPERIAL H/T (1180 + BLT., 1951-52)

51-52 (CONT'D.)

IMPERIAL SEDAN PROD. (EST.)
13,600 + ('51)
8000 + ('52)

(APPROX. CR. IMPL. LIMO. PRODUCTION, 1951-52 : 235 +)

(C-53)

$6690. ('51)
$6994. ('52)

CROWN   IMPERIAL LIMOUSINE

CHRYSLER

WINDSOR 6 (C-60-1) $2442.

WINDSOR 6

CLUB CPE. (11,646 BLT.)

("TOWN and COUNTRY" WAGON in WINDSOR and NY SERIES)

$3217.→
WINDSOR DELUXE 6 (C-60-2)
CVT. (1250 BLT.)

$3653.

$3487.↘

NEW YORKER V8 ↓

NEWPORT (2525 BLT.) (C-56-1)

NEW YORKER DELUXE
NEWPORT H/T (3715 BLT.)

SEDAN (20,585 BLT.) $3293.

POWER STEER., POWER BRAKES AVAIL.

53 (new 1-PIECE WINDSHIELDS)

(RESTYLED)

(W. and NY 8-PASS. SEDANS have 139½" WB)

(NEW YORKER DELUXE V8 is C-56-2) (new)

OTHERWISE, ALL EXCEPT IMPERIAL have 125½" WB (THROUGH '54)

CUSTOM IMPERIAL V8 (C-58) new 133½" WB

CROWN IMPERIAL V8 (C-59) 145½" WB

Imperial

BY CHRYSLER

↑
IMPERIAL (STYLIZED EAGLE) HOOD ORNAMENT (new)

$4225. TO 6872.

FOR 1954 TO 1965 IMPERIALS, SEE: IMPERIAL

CHRYSLER

8 PASS.
SEDAN
139½"
W.B.
(500
BLT.)

WINDSOR DELUXE 6
(C-62)
264.5 CID
119 HP
@ 3600 RPM
(3655 BLT.)

NEWPORT
H/T

↖ $2831.

NY 331.5 CID
V8s have
195 OR
235 HP
@ 4400 RPM

NEW YORKER
(C-63-1)

WINDSOR
DELUXE TOWN and
COUNTRY WAGON 6
$3321. (650 BLT.)

*3707.
NEWPORT
(4814 BLT.)

NEW
YORKER
DELUXE
(note SMALL
EXTRA
HORIZONTAL
CHROME PIECE
ON REAR
FENDER)

7.60 x 15 TIRES
NEW
YORKER
DELUXE V8 (C-63-2)

SEE ALSO "IMPERIAL" SECTION

54
(101,745 TOTAL SOLD)
new GRILLE
(FINAL 6-CYL.)

DASH

---

new 126" W.B. ON ALL 1955 CHRYSLERS

NASSAU OR
NEWPORT
H/Ts

← WINDSOR DELUXE (C-67) →
301 CID V8, 188 HP
@ 4400 RPM
"SPITFIRE" ENG.
(7.60 x 15 TIRES)

$2660.

4-DR. SEDAN
(63,896 BLT.)

(WINDSOR DELUXE
and N.Y. DELUXE
SERIES ONLY)

55
(RESTYLED)

N.Y. DELUXE ST. REGIS H/T
(11,076 BLT.)

$3690.

TOWN and COUNTRY WAGON (N.Y. DLX.)
(1036 BLT.)
$4209.

NEW YORKER
DELUXE (C-68)
331 CID V8,
250 HP
@ 4600 RPM
"FIREPOWER" ENG.

NEWPORT H/T
(5777 BLT.)
$3652.

(TOTAL PROD. '55
142,776)

EMBLEM
300

"300" has
300 HP @ 5200 RPM
126" WB

(1725 BLT.)

$4109.
new 300
(C-300) 331 CID

CHRYSLER

New Pushbutton PowerFlite!
(Illustrated at right)

$3336.

(C-71)
WINDSOR

CVT.
(1011 BLT.)

SEDAN
(53,119 BLT.)

DASH

$3041.

WINDSOR NEWPORT

(10,800 BLT.)

$2870.
new GRILLES

4-DR. H/Ts
NOW AVAIL.

56

NEW *"PowerStyle"* CHRYSLER

"DELUXE" DESIGNATIONS
DISCONTINUED

126"
WB

WINDSOR *has* 331.1 CID V8
(225 HP @ 4400 RPM OR
250 @ 4600)

N.Y. *has* NEW 354 CID V8
(280 HP @ 4600 RPM)

N.Y. ST. REGIS H/T
(6686 BLT.)
$3995.

(C-72)
NEW YORKER

$4523.

(1070 BLT.)

N.Y.
TOWN and
COUNTRY

NEWPORT H/T $3951.
(4115 BLT.)

MORE HORIZONTAL PCS. IN
NEW YORKER GRILLE
NOW DIFFERENT FROM
OTHERS.

BELOW:
300-B (C-72-300)
(1102 BLT.)

354 CID V8
(340 OR 355 HP
@ 5200)

NEW
YORKER
INTRODUCES
VERTICAL CHROME
STRIPS ON REAR
FENDER
(THROUGH '62.)

CHRYSLER

SEDAN $3088.

(17,639 BLT.)

WINDSOR ← (C-75-1)

$3575.

WINDSOR TOWN and COUNTRY WAGON

(2035 BLT.)

2-DR. H/T

(14,027 BLT.) $3153.

4 HEADLIGHTS ON MOST MODELS

(11,586 BLT.) 4-DR. H/T

$3832.

(C-75-2) SARATOGA

TORQUEFLITE AUTO. TRANS. AND POWER STEER. ARE STD. EQUIPMNT. ON ALL MODEL SERIES EXC. WINDSOR. (OPTIONAL ON WINDSOR AT EXTRA COST)

WINDSOR and SARATOGA HAVE 354 CID V8 (285 OR 295 HP @ 4600 RPM)

SARATOGA 4-DR. SEDAN (14,977 BLT.)

$3718.

SEDAN

note ONLY 2 HEADLTS.

(C-76) NEW YORKER

(8863 BLT.)

2 DR. H/T $4202.

57 (TOTALLY RESTYLED)

126" W.B. CONT'D. ON ALL

4-DR. H/T

$4929.

(1918 BLT.)

H/T

NEW YORKER has new 392 CID V8 (325 HP @ 4600 RPM)

4 DR. H/T $4259.

(10,948 BLT.)

300-C (C-76-300)

300-C has new HIGH and NARROWER GRILLE, also new 392 CID V8 (9.25 OR 10 COMPR.) TWO 4-BBL. CARBS.

CVT. (484 BLT.) $5359.

THE MIGHTY CHRYSLER

300/C

300-C ENGINE 375 HP @ 5200 OR 390 HP @ 5400 RPM

America's Most Powerful Car!

CHRYSLER

$5173. (6/8 BLT.)

300-D (LC3-S) →
380 OR 390 HP @ 5200 RPM

$3616.

WINDSOR
T+C WAGON — 791 6-PASS. PROD. / 862 9-PASS.

58

WINDSOR (LC1-L)

126" WB
(new SHORTER 122" WB ON WINDSOR)

EXTRA!
Now available on all Chryslers and Imperials!
AMAZING NEW **auto-pilot**
...the remarkable new device that patrols your speed—warns you when you go too fast—lets you cruise "accelerator-free"—saves gas.
ANOTHER CHRYSLER ENGINEERING EXCLUSIVE

WINDSOR *has* 290 HP @ 4600 RPM (354 CID)

WINDSOR **DART**LINE

note DIFFERENT SIDE TRIM on '58½ "DARTLINE" (ABOVE)

(ONLY 2 WINDSOR CONVERTIBLES BLT.)

$4347.

N.Y. 2 DR. H/T
(3205 BLT.)

SARATOGA
(LC2-M)

$3955.

4 DR. H/T
(5322 BLT.)

NEW YORKER
(LC3-H)
345 HP @ 4600 RPM (SAME SIZE V8 [392 CID] as 300-D)

310 HP @ 4600 RPM (354 CID)

new ENGINES: 383 OR 413 CID

CHRYSLER

59
MC SERIES

305, 325, 350 HP @ 4600, OR 380 HP @ 5000 RPM

MORE 1959 CHRYSLERS ON NEXT PAGE

CHRYSLER

#512 2-DR. H/T
(6775 BLT.)

(MC2-M)

SARATOGA

#534 4-DR. H/T
(4943 BLT.)

$4104.

(MC1-L)
WINDSOR

$3289.

LION-HEARTED
CHRYSLER '59

59
(CONT'D.)

(MC3-H)

#514 4-DR. H/T (6084 BLT.)

$3353.

#578-6 PASS. (444 BLT.)   #579-9 PASS. (564 BLT.)
N.Y.
TOWN and COUNTRY
WAGON

NEW
YORKER

#554 N.Y. 4 DR. H/T
(4805 BLT.)

$4633.

#555 N.Y. CVT. (286 BLT.)

$4890.

CHRYSLER 300 (REAR FENDER BAND)

300 E

The international classic   ...made in America

300-E
(MC3-H)

#592 2-DR. H/T
(550 BLT.)

#595 CONVERTIBLE
(140 BLT.)

$5749.

300-E
$5319.

300-E

CHRYSLER

$4067. 4-DR. H/T

WINDSOR CVT.

SARATOGA (PC2-M) *has* 383 CID V8 (325 HP @ 4600 RPM)

60

$3623. →

has 383 CID V8 (305 HP @ 4600 RPM)

SAR. SEDAN $3929 →

SARATOGA *has* GRILLE LIKE WINDSOR (ABOVE)

NEW YORKER (PC-3-H *has* 413 CID V8 (350 HP @ 4600 RPM

H/T $4461.

NEW PUSHBUTTON DASH PUTS ALL THE CONTROLS AT YOUR FINGERTIPS

NEW YORKER TOWN and COUNTRY WAGON

WINDSOR T+C $3733 UP ↗

WNDSR. →

NY CVT. **$4875.**

300/F BY CHRYSLER

The 300F medallion is molded like a gear wheel to express the rugged spirit of the car.

#23 H/T (964 BLT.)

The open grille gives the 300F a "Pure automobile" look.

413 CID V8 (375 HP @ 5000 OR 400 HP @ 5200 RPM)

→

(300-F)

$5411. (PC3-H)

#27 300 F CVT. ALSO (248 BLT.) $5841.

WINDSOR PRODUCTION:		
#23 H/T	6496 BLT.
27 CVT.	1467
41 SEDAN	25,152
43 4 DR. H/T	5897
46 6-PASS. WAGON	1120
9-PASS. "	1026

SARATOGA PRODUCTION:	
H/T	2963 BLT.
SEDAN	8463
4 DR. H/T	4099

N.Y. PROD.:		
H/T	2835
CVT.	556
SEDAN	9079
4 DR. H/T	5625
WAGON (6-PASS.)-624	(9 PASS.)-671

CHRYSLER

wagon NEWPORT T+C (2403 BLT.)

$3541. UP

(RC2-M) FINAL 1961 WINDSOR MODEL

$3303.

new NEWPORT LOW-PRICED SERIES 122" WB (RC1-L)

$3025. (NPT. H/T)

NEWPORT has new 361 CID V8 (265 HP @ 4400 RPM) (OPTIONAL 413 CID V8 has 350 HP @ 4600 RPM)

1961

4 DR. H/T (7789 BLT.) $3104.

61 (126" WB ON NEW YORKER and 300-G)

new GRILLES, CANTED HEADLIGHTS

300/G

MODELS: RC-1-L NEWPORT
RC-2-M WINDSOR (FINAL YR.)
RC-3-H NEW YORKER
RC-4-P 300-G

413 CID V8 with 350 HP @ 4600 RPM IN RC3-H NEW YORKER

NY TOWN and COUNTRY

NY

(FRONT END OF 300-G (CLOSE-UP) ILLUSTR. ON NEXT PAGE)

$4133. CHRYSLER

NEW YORKER SEDAN

CHRYSLER

300-G new GRILLE CLOSE-UP

300-G (RC4-P) has SAME ENGINES AS IN 1960

FINAL YR. of 126" WB FOR 300 SERIES

61 → (CONT'D.)

300-G
- #842 2-DR. H/T (1280 BLT.) **$5411.**
- #845 CONVERTIBLE (337 BLT.) **$5841.**

note: STARTING 1962, THERE ARE 2 300 SERIES. THE 300 CVTS. and 2-DR. H/Ts FOLLOWED BY A LETTER DESIGNATION (300-H) HAVE MORE POWER

(SCI-2) NEWPORT SEDAN (54,813 BLT.) $2964.

NEWPORT

361, 383, 413 OR new 426 CID V8 ENGINES

(10,030 BLT.) $3400.

62

300

(435 BLT.) $5090.

CVT. (123 BLT.) $5461.

265 HP @ 4400 RPM TO 421 HP @ 5400 RPM

300-H (SC2-M)

N.Y.

$4263. (6646 BLT.)

N.Y. 4-DR. H/T

$4125. SEDAN (12,056 BLT.) 126" WB

NEW YORKER (SC3-H)

ALL 122" WB (EXCEPT NY)

CHRYSLER

$3106.

ALL MODELS NOW have 122" WB. (THROUGH '64)

NEWPORT (TC1-L)

PAINTED IN ACRYLIC ENAMELS

ROUND TAIL-LIGHTS IN 1963.

$2964.

63 TC SERIES

(RESTYLED IN new "KNIFE-EDGE" [CREASE] BODY DESIGN.)

PACE CAR AT 1963 INDY 500 RACE IS 300-J.

SAME 4 V8 SIZES AS IN 1962, BUT TOP "300" HP FIGURE NOW IS 425 @ 5600, with new TOP 13.5 COMP.

DASH

(TC3-H)
NEW YORKER

#884 SALON (4 DR. H/T) (593 BLT.) (INTRO. 2-14-63)

$5860.
(14,884 BLT.) (BELOW)

NY
#878 TOWN and COUNTRY 6-PASS. WAGON (950 BLT.)

$4708.   #879 9-PASS. (1244 BLT.)
         $4815.

(TC3-H)
1963 NEW YORKERS have VERTICAL LOUVRES ON FRONT FENDERS.

#833 NEW YORKER SEDAN $4981.

NEW YORKER

CHROME BANDS JOIN ENDS of GRILLE WITH EDGES of HOOD. (NY and 300)

300

(TC2-M)
300

300-J ALSO

CHRYSLER

#814 4 DR. H/T
(9710 BLT.)

$3042.

#813 SEDAN
(55,957 BLT.)
$2901.

6 OR 9-PASS.

Chrysler Newport

Hardtop Town & Country Wagon

NEWPORT

Chrysler Newport Convertible

(VC1-L)

COMPRESSION RATIOS
NOW RUN FROM
9.0 TO 10.1
TO 1.

VC1
SERIES

64

new GRILLES
new HEADLIGHT
TREATMENT ON
NEWPORT, N.Y.

361, 383 OR 413 CID V8s
(265 HP @ 4400 RPM
TO 290 HP @ 4800)

NEW YORKER
(VC1-H)

NY SALON

VINYL TRIM
ON ROOF

note
GRILLE and
SIDE TRIM
VARIATIONS BETWEEN
"300" CVT. and H/T
MODELS
ILLUSTRATED

#833 SEDAN $3994.
(15,443 BLT.)

300 (K)

WAGON
6 OR 9
PASS.

$4721.
UP

300
(VC1-M)

INTERIOR 300

114

Chrysler Newport Convertible

(AC1-L)   NEWPORT   7-W. SEDAN

↑ 5-W. SEDAN

CVT. (3192 BLT.) $3442.

REAR INTERIOR (7-W. N.P. SEDAN)

N.Y.

NEWPORT CVT. (SHOWING DASH)

CHRYSLER MOTORS CORPORATION
CHRYSLER DIVISION

65 AC1 SERIES

NEW YORKER (AC1-H)

C-32 H/T (9357 BLT.) $4161.

C-24 4-DR. H/T (12,452 BLT.) $**4061.**

300 (AC1-M)

300 ↑

note DIFFERENCES IN SIDE TRIM BETWEEN 300 and 300-L.

C-42 H/T (2405 BLT.) $4153.

300-L

5's ...LY ...NG. ...OICES ...RE ...93 OR ...3 CID ...8s ...270 HP ... 4400 TO ...60 @ ...4800)

NEWPT. PRICES START AT $3442.

300 has ...ARGE RED ...ROSS IN CENTER OF GRILLE →

C-45 CVT. (440 BLT.)

$4618. (AC1-M)

300-L's   413 CID V8 has SPECIAL CAM.
$**4716.**
WEST COAST (CVT.)

CHRYSLER

$4086.UP
Town & Country Wagon

NEW YORKER
$4157.

440 CID V8 (350 or 375 HP)

300 →

383 CID V8 (290 or 330 HP)

PROD. 264,848

66

NEWPORT H/T $3112.

WRAP-AROUND TAIL LTS.

67

You can tell a 300 by its dash.

300 GRILLE ↓

NEWPORT GRILLE

NEW YORKER GRILLE

$3159.~4369. PRICE RANGE

1967 PROD.: 218,742

SLOT-TYPE TAIL LTS.

300

$3936.

NEW YORKER $4500.

300 has NEW CONCEALED HEADLIGHTS.

300

$4010.

1968 PROD.: 264,853

68

NEWPORT GRILLE

1968 MODELS EASILY RECOGNIZED BY NEW SMALL SIDE SAFETY LIGHTS REQUIRED BY U.S. GOV'T. REGULATIONS IN '68.

$3730.

NEWPORT CUSTOM

4-DR. H/T $4568. (WEST)

1969 PROD.: 260,773
ENGINES: 383 CID V8 (290 or 330 HP)
440 CID V8 (350 or 375 HP)

300 H/T $4104.

$3414.~4669. PRICE RANGE

69

RESTYLED

NEW YORKER GRILLE

1969

116

COMET (compact)

LINCOLN-MERCURY DIVISION *Ford Motor Company*

(INTRO. 3-60)

(TOTAL '60 SALES: 116,000)

FROM $1998. (2-DR.) **60** (*new*)

6 CYL. OHV   114" WB   90 HP

(198,021 BLT.)

two- and four-door wagons

(109½" WB)

COMET

DASH SIMILAR TO 1960

new FRONT FENDER TRIM   **61**   *new* GRILLE

FROM 2000.   "Comet" NAME MOVED TO REAR FENDER

PRODUCTION: 2-DR. (71,563); 4-DR. (85,332); 2-DR. WAGON (4,199);
4-DR. WAGON (22,165); "S-22" 2-DR. (14,004)

PRODUCTION: 144,886

62

4 DR. WAGON (16,759 BLT.)
STD. $2439.
CUSTOM: $2526.

new ROUND TAIL-LIGHTS

STD. and CUSTOM SERIES STARTS 1962
(73,880 2-DRS. BLT.)

DR.: 2084.
CUSTOM $2170.

2-DR. WAGON →

CUSTOM

CUSTOM "VILLAGER" (WOODGRAIN)

$2483.   (2318 BLT.) $2710.

S-22

4 DR. (70,227 BLT.)
$2139.
CUSTOM: $2226.

NAME RETURNS TO FRONT FENDER
new GRILLE

$2368.

117

Comet

#71-B CUSTOM
4-DR. WAGON
$2570.

new GRILLE

63

#71-C VILLAGER
4 DR. WAGON
$2754.

PRODUCTION
2 DR. (24,351); CUSTOM 2 DR. (11,897);
S-22 2-DR. (6303); CUSTOM H/T (9432);
S-22 H/T (5807); 2 DR. WAGON (623);
CUSTOM 2-DR. WAGON (272); 4 DR. WAGON (4419);
CUSTOM 4-DR. WAGON (5151);
VILLAGER 4-DR. " (1529);
(CONT'D. BELOW)

1963 PRODUCTION :
150,694

COMET *SPORTSTER* hardtop

4 DR. SEDAN (24,230);
CUSTOM 4-DR. SEDAN (27,498);
" CONVERTIBLE (7354);
S-22 " (5757)

#76-A CUSTOM CVT.
$2557.

144.3 cid
6 (85 HP);
170 cid 6
(101 HP);
221 cid V8
(145 HP); OR
260 cid V8 (164 HP)

tach, bucket seats,

Vinyl covered roof optional.

114" W.B.

DASH
(CYCLONE)

THE COMET CYCLONE.
Super 289 cu. in. V-8,
chrome engine parts,
competition-type
wheel covers.
210 H.P.
(7454 BLT.)
$2655.

(MIDSEASON
MODEL)

64
new
"ELECTRIC SHAVER STYLE"
GRILLE

$2636.

#25 CVT.
(9039 BLT.)

CALIENTE # 23 H/T
(31,204 BLT.)

PROD. :
195,227

4-DR. SEDANS
202 $2182.
404 $2269.

CALIENTE
$2350.

$2375.

Comet

(12,347 BLT.)
$2683.

#29 CALIENTE H/T

#27 CYCLONE H/T
$2578.

$2403.

(29,247 BLT.)

65

#34 CUSTOM WAGON (5226 BLT.)

1965

404

$2762. #36 VILLAGER

1965

ENJOY FAIRBANKS

200 CID 6 (120 HP) OR
289 CID V8 (200 OR 225 HP)

202
$2154.

(32,425 BLT.)

REAR FENDER DETAIL

UP-9357

40 days
from Cape Horn to Fairbanks

1965 PRODUCTION:
162,335
$2154. ~ 2762.
PRICE RANGE

CALIENTE INTERIOR →

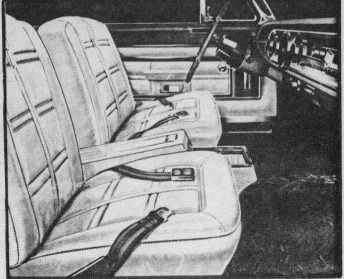

COMET 404 INTERIOR ←

COMET 404 BUCKET SEATS

COMET

1966 PRODUCTION:
133,165

BELOW: CAPRI H/T (CUSTOM SPTS. CPE)
(15,031 BLT.)

COMET *Custom Sports-Coupe*

$2400.

There are 13 models
convertibles, wagons, hardtops, sedans.

FORD MERCURY
LINCOLN

new 116" WB
(WAGONS 113")
$2475.

(25,862)

Completely equipped with white-
walls, deluxe wheel covers,
vinyl interiors, wall-to-wall
carpeting, heater-defroster,
seat belts (front and rear),
emergency flasher, lots more.

$2908.
(WEST)

Comet Caliente

$2735.

$3168.
(WEST)

(3922 BLT.)

$3152.

1966 Performance Car of the year SUPER STOCK

"Performance Car
of the Year"
Named Pace Car For
Memorial Day 500

Official PACE CAR · Mercury COMET · INDIANAPOLIS 500
CYCLONE GT

(2158 BLT.)

6.95/7.35/7.75 x 14 TIRES

66

new
GRILLES

$3510.
(WEST)

200 CID 6 (120 HP)
289 CID V8 (200 HP)
390 CID V8s
(265, 275 OR
335 HP) OR
427 CID V8

Cyclone
GT

has BODY
STRIPES

OPT.
DUAL
HOOD
SCOOPS

$2700.
CYCLONE
H/T
$3028.
(WEST)

(6889 BLT.)

$3250.
(WEST)

A.K.A.
Mercury COMET

"Performance Car of the Year"

1966 PRODUCTION: 170,426
CALENDAR YEAR: 133,165

$2154.~3152. PRICE RANGE

(13,812 BLT.)
$2891.

120

MODELS: **COMET** HORIZ.-GROOVED DASH →
02, CAPRI, CALIENTE, **CALIENTE**
CYCLONE, CYCLONE GT
$2535.
2994.
(WEST)

Caliente Grandé interior has blue Gossamer nylon or Chambrey nylon in black or parchment. Both framed with crinkle vinyl.

(INTRO. 9-30-66)

67

CYCLONE GT $3034.
(3419 BLT.)

PROD.: 56,451

3290.

CYCLONE GT   (INTRO. 9-22-67)
(RESTYLED)

68

NOTE new SIDE SAFETY LTS.

new MONTEGO

116" WB
200 CID 6
(115 HP) OR
289 CID V8 (195 HP)
MONT./COMET PROD.: 149,391

(16,693 COMETS BLT.)

WHEEL COVER

MONTEGO H/T (17,785 BLT.)
$2605.

(COMET H/T $2532.
14,104
BLT.)

69

(14,104 COMETS BLT.)

STD. ENGINES:

new 250 CID 6 (155 HP)
OR 302 CID V8
(220 HP)

V8s UP TO
428 CID

(INTRO. 9-27-68)

1969
COMET and
MONTEGO SHARE
THIS GRILLE.

corvair

GENERAL MOTORS
(1960 – 1969)

COMPACT CAR
by Chevrolet

$**1984.** and up

60

WITH THE ENGINE IN THE REAR

569 SEDAN

500
(NO CHROME BELT TRIM)

CLUB COUPE and INTERIOR (727)

700

AIR- COOLED 6- CYL.
REAR ENGINE – TRANSAXLE UNIT
140 CID
80 HP @ 4400 RPM
6.50 x 13
TIRES 108" WB

PRODUCTION =			PRICE =
500	527 COUPE	(14,628)	$1984.
	569 4-DR.	(47,683)	2038.
700	727 COUPE	(36,562)	2049.
	769 4-DR.	(139,208)	2103.
900 MONZA	927 COUPE	(11,926)	2238.

900 SERIES
(# 927)
MONZA CPE.
has DELUXE
INTERIOR
and BUCKET
SEATS.

4 DR. SEDAN and INTER.

BACK SEAT FOLDS, FOR CARGO.

$**2103.**
(769 SEDAN)

corvair

500

CLUB COUPE

$1920.

700

$2039.

700 INTERIOR

spunkier 145-cu.-in. air-cooled rear engine

4-DOOR SEDANS

$2201.

new OPTION. ELECTRIC HOT AIR HEATER

MONZA 900 #969 SEDAN $2201.

note UNIQUE WHEEL COVERS ON NEW MONZA →

new CORVAIR MONZA CLUB COUPE and INTERIOR →

61

2 new WAGON TYPES and 2 SUB-TYPES

CORVAIR GREENBRIER SPORTS WAGON

SWINGING SIDE DOORS 95" WB

$2651.

GREENBRIER (STD.)

2331.

700 (735)

LAKEWOOD 500 (535)

$2266.

LAKEWOOD STATION WAGONS

SMART, DURABLE INTERIORS—Shown here: the 700's rich fabric-vinyl upholstery, offered in three color-keyed choices. 500 all-vinyl interior also comes in three color-keyed blends. Check the push-button locks on rear doors.

700

ENGINE UNDER REAR FLOOR.

735 LAKEWOOD INTERIOR

PRODUCTION (CARS)				
527	COUPE	(16,857)	769 4-DR. SEDAN	(51,948)
535	LAKEWOOD WAGON	(5591)	927 MONZA COUPE	(109,945)
569	4-DR. SEDAN	(18,752)	969 (MONZA) 4 DR.	(33,745)
727	COUPE	(24,786)		
735	LAKEWOOD WAGON	(20,451)		

corvair

$1992.

62

500

527 COUPE
(16,245 BLT.)

FLAT 6
(80, 98 or *new*
150 HP *in* SPYDER)
(THROUGH
1963)

124130

700 $2057. UP

727 COUPE   (18,474 BLT.)
735 LAKEWOOD (3716 ")
769 SEDAN   (35,368 ")

FINAL YEAR FOR
LAKEWOOD
WAGONS

GREENBRIER
(AVAILABLE
THROUGH '65)

969 MONZA SEDAN
(48,059 BLT.)

$2273.

MONZA 900

STD. 927
MONZA CPE. IS
BEST SELLING
62 CORVAIR
(144,844
BLT.)

$2636.

MONZA SPYDER COUPE
(6894 BLT.)

CORVAIR SPYDER
(150 HP)
145 CID
FLAT
6

DASH (ALL BUT SPYDER)

63

note
CHANGE IN
FRONTAL
STYLING

MONZA

1941-X

SEDAN

JD·3668

408 62

BA·3515
WATER WONDERLAND

500   FROM   **$1992.**
700   FROM   2056.
900 MONZA   FROM   2272.
900 MONZA SPYDER   FROM   2589.

corvair

STD. ENGINE RAISED TO 95 HP.

DASH

FINAL **700**
#769 SEDAN (16,295 BLT.)
$2119.

927 COUPE
(88,440 BLT.)

64

new FRONT MEDALLION

$2281.

MONZA

SEDAN
(21,926 BLT.)
$2335.

MONZA SPYDER (ABOVE) has 150 HP.

$3008. (4761 BLT.)
(667 CVT.)

ENGINE ENLARGED TO 164 C/D
(95, 110 or 150 HP)
(LONGER STROKE)

500

←DASH has CIRCULAR GAUGES.
$2066.

H/T
(88, 954 BLT.)
$2347.

#10137 H/T
(36,747 BLT.)

MONZA

$2493.

#10567 CVT. (26,466 BLT.)

new LARGER BODIES

This year, <u>all</u> the coupes and sedans have hardtop styling

FROM
$2281.
WEST COAST

65

(ONLY MAJOR CORVAIR RESTYLING)
new 5-DIGIT BODY MODEL NOS.

New power choices, too. There's a new 140-hp engine that's standard in Corsa models and can be ordered for all others—and a 180-hp power plant that you can specify for your Corsa.

#10539 (37,157 BLT.)

MONZA SPORT SEDAN

$2422.

140 HP or 180 HP
(CORSA IS new TOP OF LINE MODEL.)

TOTAL PRODUCTION, 1964 = 195,780
" " 1965 = 204,007

CORVAIR

MONZA

$2350.

A most unusual car for people who enjoy the unusual

H/T $2556. (WEST COAST)

(WEST COAST) $2630.

SPT. SEDAN (4-DR. H/T)

CORSA $2662.

1966 SALES: 88,951

monza

66

108" WB   6 CYL., 164 CID
7.00 × 13 TIRES   95-140 HP

$2424.

CORSA SERIES NO LONGER AVAILABLE 1967

95 OR 110 HP ENGINES ONLY DURING 1967.

'67 Corvair
The rear-engine road car

1967 SALES: 24,736

500 = 2 DR. H/T	$2128.	
4 DR. H/T	2194.	
MONZA = 2 DR. H/T	2398.	
4 DR. H/T	2464.	
CONVERT.	2540.	

500

('67)

THIS BEST IDENTIFIES A 1967 MODEL.

New oval steering wheel—This easy-to-grip wheel sits atop the GM-developed energy-absorbing steering column—one of many new standard safety features. Others include 4-way hazard warning flasher and a lane-change feature incorporated in direction signal control.

67

142D 91

('67)

MONZA CVT.

new DASH

OPTIONAL LUGGAGE RACK

CORVAIR

(NO 4-DR. HARDTOPS AFTER 1967)

new SIDE SAFETY LIGHTS

Corvair 500
H/T
$2528.

68-69

SALES:
12,977 (1968)
3102 (1969)

DISCONTINUED MAY 14, 1969

$2641.

MONZA CONVERTIBLE

1969 PRICES SHOWN

MONZA SPT. CPE.
$2522.

126

(STARTS 1953)

CORVETTE

(315 BLT. 1953, 3640 BLT. 1954)

MODEL 290 (UNTIL '57) (A.K.A. #2934.)

102" WB (THROUGH '62)   235½ CID 6 CYL. CHEVROLET ENGINE

53 -54
$3512. ('53)
3523. ('54)

CHEVROLET Corvette

INTERIOR (BROCHURE ILLUSTR.)

FULL-LENGTH SIDE TRIM on REGULAR-PRODUCTION

Sports Car

FIBERGLASS BODIES on ALL (TO DATE)

6 CYL. or new V8 (265 CID, 162 HP)

55

ILLUSTRATED with AVAIL. DETACHABLE TOP

(674 BLT., 1955)

(BECAUSE OF COMPETITION with FORD's new THUNDERBIRD,) 1955 CORVETTE PRICE CUT TO $2799.

V8 ENGINE ONLY, 1956 ON (265 CID, 1956; 283 CID, 1957 THROUGH '61)

H.P. IN 1957 =
220 (STD.)
245-270 (OPT.)   250-283 w. FUEL INJECTION

new TOP

SINCE '56, new SIDE TRIM

56-57
$2900. ('56)

$3437. ('57)

3467 BLT., 1956)

new REAR FENDERS   (6339 BLT., 1957)

102" WB
230 HP

DASH

new 4 HEADLIGHTS

58
(9/68 BLT.)

RESTYLED

new VENT LOUVRE GROUP ON TOP OF HOOD (1958 ONLY)

$3631.
new BUMPERS

MODEL 867 (THROUGH 1964)

230 HP (STD.) 245-270 (OPT.)
250-290 (1959 w. FUEL INJ.)
275-315 (1960 w. FUEL INJ.)

59-60

(10,261 BLT., 1960)

$3872.
(IN '60; $3 LESS THAN '59)

9670 BLT., 1959)

XX 8956

127

CORVETTE

$4272. WEST COAST **61** $3934.

SAME H.P. CHOICES AS in 1960

(10,939 BLT.) new GRILLE

new 250 HP STD. new 300-340 or 360 HP (F.I.) OPTIONAL (THROUGH '63) **62** (14,531 BLT.) **$4375.** WEST COAST $4038.

new SIDE-SCOOP DESIGN

63 new "STINGRAY"

new SIDE-SCOOPS AGAIN

new GRILLE, CONCEALED HEADLIGHTS, new 98" WB

$4589. WEST COAST new #837 SPORTS COUPE (10,594 BLT.) FASTBACK $4252. new SHORTER 98" W.B. #867 CONVT. RDSTR. (10,919 BLT.) $4037.

H.P. 250 (STD.) 300 (OPT.) 395 (with FUEL INJECTION)

(8304 BLT.) (13,925 BLT.) new 1-PC. BACKLIGHT → new 1-PC. **64**

Corvette Sting Ray Convertible in Saddle Tan
Corvette Sting Ray Sport Coupe in Riverside Red

$4252. $4037. WEST COAST $4627.

WEST COAST $4723. 327 CID V-8 has 250, 300, 350, 365 or 375 HP @ 5500 RPM **65** W. ADD-ON TOP WEST COAST $4508.

#19437 SPT. CPE. (8186 BLT.) **4-WHEEL DISC BRAKES** $4321. #19467 CVT. RDS. (15,376 BLT.)

$4106. new VERTICAL LOUVRE DESIGN

425 HP 396 CID V8 →

1965½ CORVETTE "396"

Corvette

98" WB
(SINCE '63)

7.75 x 14
TIRES

327 cid V8 (300 or 350 HP)
427 cid V8 (390 HP)
(OPT. 425 HP, '66 ; 435 HP, '67)

66-67

CONVERTIBLE AVAIL. ALSO, AT
$4084. ('66) $4141. ('67)
17,762 BLT.   14,436 BLT.

9958 BLT. '66
8504 BLT. '67

COUPE
$4295. ('66)
4353. ('67)

COUPE
(9936 BLT.)

4663.

WEST COAST:
$5157.
CPE.

new F70 x 15 TIRES

1968 TOTALLY RESTYLED

$4320.

OPTIONAL
HARD
TOP

(18,630 BLT.)

CVT.
$4814.
(WEST COAST)

new T-TOP (INTERIOR)

Corvette simulated wood steering wheel and instrumentation.

'68

4781.
22,154
(BLT.)

COUPE (22,154 BLT.)
$4781.

CONV'T.
$4438.
(18,608
BLT.)

69

"STINGRAY"
NAME ADDED,
ON FRONT FENDERS

327 cid REPL. BY NEW
350 cid V8
(300 HP)
427 cid V8 CONT'D.
(390, 400, 430
or 435 HP)

EU 9262

Cougar (STARTS 1967)
BY MERCURY

$2851.
$3213. (WEST COAST)

H/T (116,260 BLT.)

New
(LINCOLN/MERCURY'S COUNTERPART TO FORD'S POPULAR MUSTANG)

111" WB
289 CID V8 (200 HP)
7.35 × 14 TIRES
390 CID V8 (320 HP) IN "GT PERFORMANCE GRP."

67

COUGAR ADVERTISING MASCOT

(GT H/T 7412 BLT. $3175.)

WITH TRIP ODOMETER

XR-7 $3081.
$3443. (WEST COAST)

has GRAINED DASH and BEARS THIS SYMBOL

(27,221 BLT.)

note EMBLEM ON HEADLIGHT COVER SECTION

TOTAL 1967 PRODUCTION: 150,893

CALENDAR YEAR PRODUCTION: 131,743

COUGAR

$2933.

E 70 × 14 TIRES
289 CID V8 STD.
427 CID V8 (390 HP)
IN GT. E

(GT MODEL ALSO)

electric sunroof

$3232.
(32,712 BLT.)

$3296.
(WEST COAST)
(81,014 BLT.)

68

$3594.
(WEST COAST)

XR-7-G has
SPORT-STYLE HOOD
and RALLYE LIGHTS
(GT.E has HORIZ. BAND ACROSS GRILLE)

new
SAFETY SIDE LIGHTS

$3016.
FROM $3383.
(WEST COAST)

1969½
"ELIMINATOR"
(not illustr.)
has
new
FRONT
and
REAR
SPOILERS.

VARIOUS V8s
AVAIL., INCLUDING
new 351 CID V8 (250 HP @ 4600 RPM)
E 78 × 14 TIRES

69

new
DOWNSWEPT
SIDE SCULPTURE

new GRILLE

(new STD. or XR-7 CONVERTIBLES ALSO
AVAIL.)

TOTAL 1969 PRODUCTION:
100,069

H/T (66,331) $3016.
XR7 H/T (23,918) $3315.

CONVERT. (5796) $3382.
XR7 CONVERT. (4024) $3595.

CROSLEY

(1939-1952)

MFD. IN MARION, IND.

39-42

2-CYL. AIR-COOLED
WAUKESHA ENGINE (35.3 CID)
(THROUGH '42)

12 HP
MECHANICAL
BRAKES

80" WB
4.25 × 12 TIRES

POWEL
CROSLEY, JR.

FOUNDER OF
CROSLEY CORP.
(KNOWN AFTER WW
2 AS CROSLEY MOTORS)

PRICE CUT TO
$299.
IN 1941.

$412.
IN 1942

CVT. (OTHER MODELS
ALSO AVAIL.)

PRODUCTION (ALL BODY TYPES): 422 (1940); 2289 (1941); 1029 (1942)

CAR PRODUCTION: CONVERTIBLES:
12 (1946); 4005 (1947); 2845 (1948)

2-DR. SEDANS:
4987 (1946); 14,090 (1947); 2750 (1948, INCLUDING
new SPORT UTILITY 2-DR.)

2-DR. WAGON: 1249 (new, 1947); 23,489 (1948)

"CC" SERIES

"a FINE car" new 4-CYL. WATER-COOLED "COBRA" (COPPER-BRAZED)
STAMPED-BLOCK 44 CID ENGINE (26½ HP @ 5400 RPM)

new BODY SIDES COMBINE
with FULL-LENGTH FENDERS

WAGON

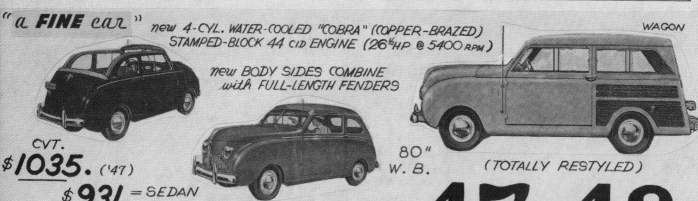

CVT.
$1035. ('47)

$931. = SEDAN
('47) new GRILLE, BUILT-IN HEADLIGHTS ABOVE

80"
W. B.

(TOTALLY RESTYLED)

47-48

(POSTWAR PRODUCTION RESUMES DURING
JUNE, 1946)

PICKUP

4.50 × 12"
TIRES

$799.
('48)

CROSLEY SPORTS-UTILITY

PANEL DELIVERY

132

CROSLEY

48½

new GRILLE
ON MID-YEAR
"NEW LOOK" SERIES

FOR 1949, COPPER-BRAZED, 58-lb. STAMPED ENGINE
REPLACED BY IMPROVED CAST-IRON VERSION (CIBA.)

new GRILLE, "SPEEDLINE"
STYLING

SEDAN

49-50

new 85" W.B.
"HOTSHOT"
SPORTS
ROADSTER

(ALSO "SUPER
SPORTS")

$866. ('49)
$882. ('50)

new
HYDRAULIC
DISC
BRAKES (BY
GOODYEAR-
HAWLEY)

CVT.

WAGON

(645 BLT. 1949)
(478 BLT. 1950)

"CD" SERIES
(THROUGH 1952)

(3803 BLT. 1949)
(4205 BLT. 1950,
INCLUDING *new*
$984 "SUPER WAGON")

ROADSTERS : 752 BLT. 1949
742 BLT. 1950

INSTRUMENT
CLUSTER
(HOTSHOT)

new BENDIX
9" HYDRAULIC
BRAKES

new "BUSINESS COUPE" VARIATION
OF 2-DR. SEDAN ($943.)

51-52

Crosley Hotshot

new 2-BLADED GRILLE *with*
CENTER "SPINNER"

"SUPER WAGON"

$952. $1029. (S.S.)
646 RDSTRS. BLT. 1951,
358 " BLT. 1952

(9500 BLT.,
1951)
(1355 BLT. 1952)
(WAGON, SU. WAGON)

$1450. ('51)
$1077. ('52)

DISCONTINUED DURING 1952

DART

Dodge Division of Chrysler Corporation

FULL-SIZED
LOWER-PRICED
new COMPANION
TO

DODGE STARTS 1960

(BECOMES A COMPACT CAR, 1963)

SENECA

$2410. (6)
$2530. (V8)

(STARTS 1960)

M 21
2 DR.

L 45 WAGON $2695. (6) $2815. (V8)

PIONEER

H 23 2-DR. H/T
$2618. (6) $2737. (V8)

PHOENIX

H 43 4-DR. H/T
$2677. (6)   2796. (V8)

118" WB (WAGONS 122")
(THROUGH '61)

60 new!

PD3 (6 CYL.)
PD4 (V8)

225 CID SLANT 6 has
145 HP @ 4000 RPM
318, 361 and 383 CID V8 have
230, 255, 310, 325
OR 330 HP.

$2787. UP

PIONEER WAGON
M 45 A (6-PASS.)
M 45 B (9-PASS.)

THE DODGE DART IS PRICED MODEL FOR
MODEL WITH OTHER LOW-PRICE CARS.

DODGE DART	CAR F	CAR P	CAR C
SENECA	Fairlane	Savoy	Biscayne
PIONEER	Fairlane 500	Belvedere	Bel Air
PHOENIX	Galaxie	Fury	Impala

PRODUCTION

SENECA 6	(93,167)
SENECA V8	(45,737)
PIONEER 6	(36,434)
PIONEER V8	(74,665)
PHOENIX 6	(6567)
PHOENIX V8	(66,608)

H27 PHOENIX
CONVERTIBLE AVAIL.,
$2868. UP

1960

(HIGHEST-PRICED MODEL = PIONEER V8 9 PASS. WAGON, $3011.)

WAGON
with
TAILGATE
OPEN

New Economy Slant "6" Uses
Exclusive Semi-Ram Intake Manifold!

New design
features
inclined block
with new
Equi-flow fuel
induction, over-
head valves,
for greater
fuel economy.

DART $2695. UP

145 HP 6-CYL. CONTINUES

L456 (6)   L556 (V8)

SENECA

M466, M467 (6)
M566 M567 (V8)

SENECA STATION WAGON 6 OR V8,
6 PASSENGER

SENECA 4 DOOR SEDAN 6 OR V8

$2330. UP

PIONEER STATION WAGON 6 OR V8,
5 OR 9 PASSENGER

PIONEER 4 DOOR SEDAN 6 OR V8

$2459. UP

RD3 (6 CYL.)
RD4 (V8)

$2787. UP

PIONEER

$2595. UP

61

318, 361, 383 and new 413 CID V8s
(230 TO 375 HP)

PHOENIX 4 DOOR HARDTOP 6 OR V8

PHOENIX

convertible

H535 CONVT. (V8 ONLY)
$2988.

PRODUCTION =		
SENECA 6 (60,527)	SENECA V8 (27,174)	FINAL YEAR
PIONEER 6 (18,214)	PIONEER V8 (39,054)	FOR THESE
PHOENIX 6 (4273)	PHOENIX V8 (34,319)	MODEL NAMES

DART 330 2-DOOR HARDTOP 6 OR V8

DART 330 4-DOOR 6-PASSENGER WAGON 6 OR V8

DART 330 2-DOOR SEDAN 6 OR V8

2463. UP

$2739. UP

$2375. UP

$3092.

DART 440 9-PASSENGER WAGON V8

new 116" WB
('62 ONLY)

SAME DISPL. AS '61
145 TO 380 HP

62 (TOTALLY RESTYLED)

DART 440 CONVERTIBLE V8 $2945.

MODELS

DART 6 = (SD1-L)
"   " 300 (SD1-M)
"   " 440 (SD1-H)
DART V8 = (SD2-L)
"   " 330 (SD2-M)
"   " 440 (SD2-H)

(BUDGET-PRICED
"FLEET SPECIAL",
$2158. UP)

1962 IS FINAL YEAR
THAT DART IS
DODGE-SIZED.

SD SERIES

THE NEW LEAN BREED OF DODGE

DART

713 SEDAN

$2041. 170 (RESTYLED)

63 TL1 SERIES

756 WAGON $2309.

731 2-DR.

8.2 COMPR.

WHEELBASE REDUCED AGAIN, TO 111" (106" WB ON WAGONS)

270

DASH

$2512.

776 WAGON $2433.

PRODUCTION
170 (58,536)
270 (61,159)
H/T GT (34,227)

745 CVT.

GT 742 H/T

ALL 1963 DARTS ARE 6-CYL.
170 CID   101 HP @ 4400 RPM
OR 225 CID 145 HP @ 4000 RPM

$2289.

$2318. #742 H/T

TWO SLANT SIXES AS IN '63
new 273 CID V8 (180 HP @ 4200)

$2315. 170

GT

64 VL1 (6) VL2 (V8) "DODGE" NAME ACROSS new GRILLE (270 STILL AVAIL.) $2053. SEDAN

$2407.

new GRILLE ENGINES AS IN 1964

65 AL1 (6) AL2 (V8)

$2481.

Dodge Dart 4-door station wagon. 2-seat model only. 6 and V8 power.
L-56

Dodge Dart 270 convertible, 6 and V8 power.

270

DART (EX-170)

GT

Dodge Dart 2-door sedan. 6 and V8 power.

Dart GT two-door hardtop.

L-11

L-42

1965 PRODUCTION (BY MODEL SERIES)
DART (AL1-L) (86,013 BLT.) $2074. UP
DART 270 (AL1-H) (78,245 BLT.) $2180. UP
DART GT (AL1-P) (45,118 BLT.) $2404. UP

L-45 (GT CONVERT. ALSO AVAILABLE, $2628.)

Dart

2-DR. $2319.

170 CID 6 (101 HP)

DART

225 CID 6 (145 HP) OR

273 CID V8 (235 HP)

Dart 4-door sedan. Six or 273 V8 power.

$2383.

$2661.

JOIN THE DODGE REBELLION

WAGONS $2094. ~ 2828. PRICE RANGE
$2319. ~ 2925. WEST COAST PRICES (LISTED HERE)

66 new GRILLE

6.50 × 13 TIRES

111" WB (106," WAGONS)

DART 270 SERIES

2-DR. $2439.

270 WAGON $2758.

270 CVT. $2795.

270 H/T $2532.

GT

270 4-DR. $2505.

GT HAS CHROME ATOP FR. FENDERS AND ON ROCKER PANELS

GT H/T $2642.
PARTIAL VINYL TOP AVAIL.

GT CVT. $2925.

TOTAL 1966 PRODUCTION = 176,027 (75,990 DARTS; 69,996 DART 270s; 30,041 DART GTs)
(146,361 CALENDAR YEAR)

$2453. **DART**

$2416.
(WEST COAST)

DART WAGONS DISCONTINUED

(INTRO. 9-29-66) **67**

(RESTYLED)

←MODELS (6 OR V8)
CLL **DART** = (53,043 BLT.) $2187. ~ 2352.
CLH **DART 270** = (63,227 ") 2362. ~ 2516.
CLP **DART GT** = (38,225 ") 2499. ~ 2860.
(WEST COAST PRICES HIGHER)

ENGINES :
170 CID SLANT-6 (115 HP)
225 " " " (145 HP)
273½ " V8 (180 OR 235 HP)

new SUNKEN-IN REAR WINDOW↘

$2591.

$2388. (6)
2516. (V8)

DART 270

$2617.
(WEST COAST)

The Dodge Rebellion: Operation '67

CONVERTIBLE (GT)
$2732. (6) 2860. (V8)

GT

Go '67 Dart!"

WEST COAST
$2961.

$2728.

←REDESIGNED RECESSED INSTRUMENT PANEL

note "GT" TAGS

FULL 1967 LINE ILLUSTRATED

DeSoto

BY CHRYSLER CORP., AVAIL.
1929 to 1961 MODEL YEARS

SUBURBAN
$2631.
('48)

7 TAXI

ALL '48-STYLE
CHRYSLER CORP. CARS
CONT'D. TO 2-49.

SINCE LATE 1935, LONG-WHEELBASE DESOTO
7-PASS. or 8-PASS. SEDANS and LIMOUSINES
AVAIL., MANY SOLD in FLEETS to BIG-CITY
TAXICAB COMPANIES
139½" LONG W.B. AVAIL.
(FROM 1940 THROUGH 1954.)

S-11S=
DELUXE
S-11C=
CUSTOM

S-11
46-48

CUSTOM

6 CYL. (236.6 cid) 109 HP @ 3600 RPM

TIRES : 6.50 × 15 , 6.50 × 16 L.W.B. ('46-47)
7.00 × 15 , 7.50 × 15 L.W.B. ('48)

SEMI-AUTOMATIC TRANS.
and "FLUID DRIVE"

CONVENTIONAL
HEADLIGHTS
RESUMED

LARGER DIE-CAST
GRILLE

$1761. ('46) $1965. ('47)

"8 out of 10
say DeSoto again*"

$2296. ('48)

*=SLOGAN BASED on POLL WHICH
INDICATED HOW MANY WOULD BUY ANOTHER DE SOTO.

1946 to FEB., 1949   DLX.= CPE. (1950); CLUB CPE. (8580); 2-DR. SEDAN (12,751); 4-DR. SEDAN (32,213)
PRODUCTION   CUSTOM = CLUB CPE. (18,431); CONVT. (3385); 4 DR. SEDAN (48,589); 8-PASS. SEDAN (342)
9-PASS. SUBURBAN SEDAN (129) (ADDITIONAL TAXI SALES, 7500 (?))

"THE CAR DESIGNED WITH YOU IN MIND"

49
(TOTALLY
RESTYLED)

S-13 {-1=DELUXE
{-2=CUSTOM
(STARTS 2-49)

6000-W

REAR

EMBLEM

new 125½" WB (THROUGH '54)

(1949 CONT'D.
NEXT PAGE)

DE SOTO CUSTOM

4 DR. SEDAN
13,148 BLT.)
$1986.

DE LUXE
(has NO EXTRA CHROME
FENDER STRIPS.)

CONVERTIBLE
(3385 BLT.)
$2578.

CLUB COUPE
(18,431 BLT.)
$2156.

49
(CONT'D.)

new 112 HP
@ 3600 RPM
(THROUGH
'50)

139½" W.B.
8 PASS. SEDAN
(AVAIL. THROUGH
'54)

CONVERTIBLE
(2900 BLT.)
$2578.

"CARRY ALL"
SEDAN
(new)
(IN LOWER-
PRICED
"DELUXE"
SERIES, BUT
HAS ADDITIONAL
CHROME TRIM.
(2690 BLT.)

$2191.

AS IN 1949,
DE LUXE PRICES
START AT
$1976. (CLUB CPE.)

RE-DESIGNED LIKENESS
of HERNANDO DE SOTO, HISTORIC
SPANISH EXPLORER FOR WHOM
CAR WAS
NAMED

'50 GRILLE has
PAINTED SECTION
IN CENTER, with
new EMBLEM.

CUSTOM

IN "CUSTOM" SERIES,
2 TYPES of 4-DR.
STATION WAGON
BODIES IN '50:
WITH WOOD
(600 BLT.)
$3093.
OR
ALL-STEEL
(100 BLT.)
$2717.

50 CUSTOM

new!
SPORTSMAN
H/T (4600 BLT.)
$2489.

REAR

SEDAN (72,664 BLT.)
$2174.

new ROUND
PARKING LIGHTS

Drive a De Soto before you decide!

DE SOTO

SPORTSMAN H/T $2761.

S-15-1 (DE LUXE)   S-15-2 (CUSTOM)

51

6-CYL. DISPLACEMENT RAISED TO 250.6 CID 116 HP @ 3600 RPM (THROUGH '54)

new LOWER, SIMPLER GRILLE

AS ON OTHER '51 CHRYSLER CORP. CARS, new "ORIFLOW" SHOCK ABSORBERS

$2436.

1951 MODELS have SCRIPT LETTERING ABOVE GRILLE

S-15 MODELS CONTINUE CUSTOM 6

SEDAN CUSTOM

new Full Power Steering

52

1952 MODELS have BLOCK LETTERING ABOVE GRILLE

WAGON

new FIRE DOME V8 (BELOW and RIGHT) (S-17)

S-17 CARS with AIR SCOOP HOOD ORNAMENT HAVE new "FireDome" 276.1 CID V8 ENGINE

$3/83. →

(850 V8 CVTS. BUILT)

Power Braking

160 h.p. @ 4000 RPM

L-HEAD 6-CYL. ENG. STILL AVAIL. (THROUGH 1954)

DESOTO-PLYMOUTH Dealers present **GROUCHO MARX** in "You Bet Your Life" every week

DASH

'51-'52 PRODUCTION : **DE LUXE** = CLUB CPE. (6100); SEDAN (13,506); CARRY-ALL (1700); 8-PASS. SED. (343); **CUSTOM** = CLUB CPE. (19,000); H/T (8750); CVT. (3950); SEDAN (88,491); WAGON (1440); 8-PASS. SED. (769); 9-PASS. SUBURBAN (600); **FIREDOME V8** = CLUB CPE. (5699); H/T (3000); SED. (35,651); WAGON (550); 8-PASS. SED. (50)

(FIREDOME V8 NOT AVAIL. 1951)

DeSoto

POWERMASTER 6 or FIREDOME V8 MODELS

new 1-PC. WINDSHIELD

$2893.

POWERMASTER 6
$2356. (S-18)

6 has a BROAD SHIELD EMBLEM on HOOD; V8 has "V" BELOW a NARROWER SHIELD (THROUGH '54)

V8 SPORTSMAN H/T

RESTYLED **53**

PRODUCTION =
CLB. CPE (6) (8063); (V8) (14,591)
H/T (6) (1470); (V8) (4700)
CVT. (V8 ONLY) (1700)
SEDAN (6) (33,644); (V8) (64,211)
WAGON (6) (500); (V8) (1100)
8-PASS. SEDAN (6) (225)
(V8) (200)

V8 CONVERTIBLE $3114.

REAR DETAILS (V8 SEDAN)

2643.

new POWER BRAKES and OVERDRIVE AVAILABLE

7.60 x 15

V8 CLUB CPE.
$2622.

THE FINAL 6-CYL. DE SOTO

POWERMASTER 6

(S-20)

"POWERFLITE" A.T. AVAIL.

V8 has new 170 HP @ 4400 RPM

CORONADO SEDAN

V8 SED.
(45,093 BLT.)
$2673.

54

FIREDOME V8 (S-19)

DASH

GRILLE MODIFIED new SIDE TRIM and TAIL-LIGHTS

143

DE SOTO — The *Forward Look*

V8s ONLY (1955 ON)

new 126" WB

(TOTALLY RESTYLED)

55

FLITE-CONTROL gear selector lever is mounted on De Soto's smart, new instrument panel—out of your way. Yet at your finger tips.

7.60 x 15

new 291 CID
FIREDOME (S-22)   185 HP @ 4400 RPM
FIREFLITE (S-21)   200 HP @ 4400 RPM

PRODUCTION: S-22 FIREDOME SPECIAL H/T; SPORTSMAN H/T (28,944); CONV'T. (625); 4-DR. SEDAN (46,388); 4-DR. WAGON (1083)
S-21 FIREFLITE SPORTSMAN H/T (10,313); CONV'T. (775) 4-DR. SEDAN; CORONADO (26,637)

PRICE RANGE = $2498. TO $3170.

DRIVE A DE SOTO BEFORE YOU DECIDE

(PUSH-BUTTON A.T.)

$3728.
new 341.4 CID
ADVENTURER (S-24)
320 HP @ 5200 RPM
(996 BLT.)

DASH (MINOR CHANGES FROM 1955)

230 HP @ 4400 RPM
FIREDOME (S-23)

"HIWAY HI-FI" BLT.-IN RECORD PLAYER AVAIL.

new MESH GRILLE

TRIPLE TAIL-LIGHTS with OVERLAPPING FIN

(44,909 BLT.)

new 330 CID

56

$2678.
PACESETTER CVT. (100 BLT.) $3615.

255 HP @ 4400 RPM
FIREFLITE (S-24)

SEDAN $3119.

(18,207 BLT.)

PACE CAR AT 1956 INDY 500 RACE

new 12-VOLT ELEC. SYS.

144

DE SOTO...the most exciting car in the world today!

PRICE RANGE =
$2777. TO $4272.

SPORTSMAN 2-DR. H/T
(13,333 BLT.)
$2836.

DE SOTO

$2912. new LOWER PRICE FIRESWEEP (S-27)
(has OWN FRONT END STYLING)

(S-25) $2958.
FIREDOME

4-DR. H/T
(7168 BLT.)

new 325, 341
OR 345 CID

245, 270, 295
OR 345 HP

SEDAN
(23,339 BLT.)

(TOTALLY RESTYLED)

57

$3614.

(1151 BLT.)

2-DR. H/T

H/T (1650 BLT.)

$3997.

ADVENTURER

(4 HEADLIGHTS, ANODIZED GOLD TRIM)
(S-26)

FIREFLITE
(S-26)

new 122" WB
ON FIRESWEEP;
126" ON OTHERS
(THROUGH '59)

$3671.

$3487.

4-DR. H/T
(6726 BLT.)

REAR SEAT FACES BACKWARD
ON EXPLORER WAGON
(3 SEATS)
9-PASS.

BOTH WAGONS ARE AVAIL. IN EITHER FIRE-SWEEP OR FIREFLITE SERIES.

SHOPPER (2 SEATS) 6-PASS.

FIREFLITE
SEDAN (11,565 BLT.)

WAGON PRODUCTION = FIRESWP. 6-PASS. (2270); 9 PASS. (1198); FIREFLT. 6 PASS. (837); 9-PASS. (934)

DE SOTO

18 MODELS, 4 SER.

new "TURBOFLASH" V8 (350 or 361 CID)

280 TO 355 HP

(LS2-M) FIREDOME $3235.

(LS1-L) FIRESWEEP $2953.

$3675.

$4172.

FIREFLITE (LS3-H)

CLOSE-UP

58

LS SERIES

9-PASS. "EXPLORER" WAGON

LARGE new "CONTROL TOWER" WINDSHIELD

DASH

GULFLEX SERVICE GULF

$4369. (LS3-S)

ADVENTURER has ANODIZED SIDE TRIM

DE SOTO – the exciting look and feel of the future!

PRODUCTION: **FIRESWEEP** H/T (5635); CVT. (700); SEDAN (7646); 4 DR. H/T (3003);
6-PASS. WAGON (1305); 9-PASS. WAGON (1125); **FIREDOME** H/T (4325); CVT. (519); SEDAN (9505);
4-DR. H/T (3130); **FIREFLITE** H/T (3284); CVT. (474); SEDAN (4192); 4-DR. H/T (3243);
6-PASS. WAGON (318); 9-PASS. WAGON (609)
ADVENTURER H/T (350); CVT. (82)

PRICE RANGES:

FIRESWEEP (MS1-L)	= $2904.	TO	$3508.
FIREDOME (MS2-M)	= $3234.	TO	$3653.
FIREFLITE (MS3-H)	= $3763.	TO	$4358.
ADVENTURER (MS3-S)	= $4427. (H/T)	OR	$4749. CVT.

'59 DE SOTO

(MS1-L) FIRESWEEP

#23 H/T $2967.

(5481 BLT.)

8.00 x 14 TIRES

361 OR new 383 CID V8 (THROUGH '60)
295 HP @ 4600 RPM
TO 350 HP @ 5000 RPM

23 SPORTSMAN H/T (1393 BLT.) $3831.

(MS3-H) FIREFLITE

41 SEDAN (4480 BLT.)

$3763.

FIREFLITE SHOPPER

(ALL BUT FIRESWEEP HAVE 8.50 x 14 TIRES.) (SINCE '57)

(MS2-M) FIREDOME

#23 SPORTSMAN H/T (2862 BLT.)

ADVENTURER (MS3-H)

#27 CVT. (97 BLT.)

DASH

147

1960 DE SOTO

#41 SEDAN
(9032 BLT.)

(PS1-L)
FIREFLITE

DASH (with RAISED INSTRUMENT CLUSTER)

4-DR. H/T
(2759 BLT.)

BUILT-IN 45-RPM RECORD PLAYER OPTIONAL AGAIN (AS IN PLYMOUTH)

(PS3-M)
ADVENTURER $3727.

H.P. CHOICES :
295 @ 4600 ; 305 @ 4600 ;
325 @ 4600 OR
330 @ 4800 RPM

ONLY 2 MODEL SERIES IN 1960:
FIREFLITE OR
ADVENTURER
(NO MORE DE SOTO WAGONS OR CONVERTIBLES)

#23 H/T
(3092 BLT.)
$3663.

10.0 TO 1 COMPRESSION

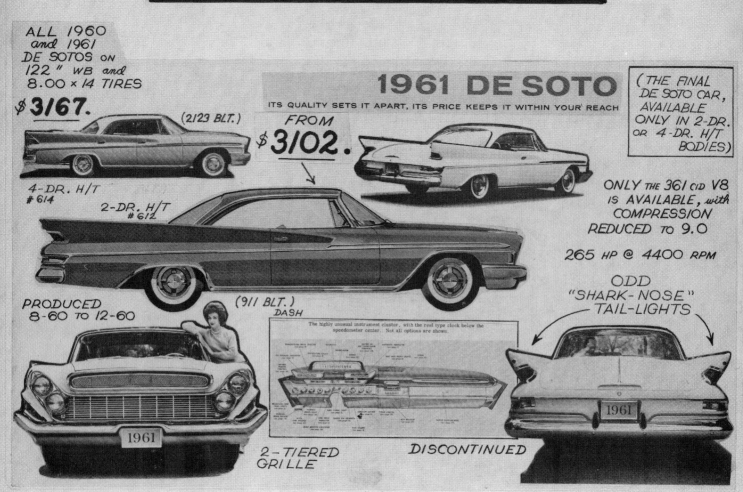

ALL 1960 and 1961 DE SOTOS ON 122" WB and 8.00 × 14 TIRES
$3167.

(2123 BLT.)

1961 DE SOTO
ITS QUALITY SETS IT APART, ITS PRICE KEEPS IT WITHIN YOUR REACH

FROM
$3102.

(THE FINAL DE SOTO CAR, AVAILABLE ONLY IN 2-DR. OR 4-DR. H/T BODIES)

4-DR. H/T
#614

2-DR. H/T
#612

ONLY THE 361 CID V8 IS AVAILABLE, with COMPRESSION REDUCED TO 9.0

265 HP @ 4400 RPM

ODD "SHARK-NOSE" TAIL-LIGHTS

PRODUCED 8-60 TO 12-60

(911 BLT.)
DASH

The highly unusual instrument cluster, with the reel type clock below the speedometer center. Not all options are shown.

2-TIERED GRILLE

DISCONTINUED

DODGE

ESTABLISHED
1914,
MFD. BY CHRYSLER
CORP. SINCE 1928

1946-48 PRODUCTION
D 24S DELUXE =
BUSINESS CP.	27,600
2 DR. SEDAN	81,399
4 DR. "	61,987

D 24C CUSTOM =
CLUB COUPE	103,800
CONVERTIBLE	9500
4 DR. SEDAN	333,911
TOWN SEDAN	27,800
7-PASS. SEDAN	3698
7-PASS. LIMO.	2
CHASSIS ONLY	302

(SERIES ENDS 2~49

(ABOVE)
2-DR. SEDAN
$1676.

CLUB COUPE
$1774.

7-WINDOW SEDAN $1788.

DASH and INTERIOR VIEWS

new
7.10 × 15
TIRES IN
1948
119½" W.B.
(137½", 7-PASS.)

CONVERTIBLE
$2189.

D-24S = DE LUXE
D-24C = CUSTOM

46-48

230.2 CID L-HEAD 6
102 HP @ 3600 RPM

$1872.

TOWN SEDAN
(5-WINDOW SEDAN)
(ALL DOORS FRONT-HINGED)

"FADEAWAY"
FENDERS

Dodge

SMOOTHEST CAR "AFLOAT"

ROADSTER (new)
(TOP UP)
$1611.

(ACTUAL PHOTO) (TOP DOWN)

3-WINDOW COUPE

2-DOOR $1738.

LOWER PRICED
NEW DODGE *WAYFARER*

The Daring New
DODGE
gyrol Fluid Drive plus GYRO-MATIC
Frees You from Shifting
OPTIONAL ON CORONET MODELS

PRODUCTION =
WAYFARER
3 - WINDOW CPE.   9342
2 - DR. SEDAN   49,058
ROADSTER   5420
MEADOWBROOK
4 - DR. SEDAN }
CORONET } 144,390
4 - DR. SEDAN }
CLUB COUPE   45,345
CONVERTIBLE   1800
4 - DR., 9 PASS.
    WAGON   800
CHASSIS ALONE   1

MEADOWBRK. 4-DOOR
SEDAN,
$1848.

MEADOWBRK.
HAS HUB CAPS
INSTEAD OF
FULL WHEEL
COVERS.

New Dodge CORONET

SEDAN $1927.

$1727.

(ABOVE) WAYFARER ROADSTER (ARTIST'S CONCEPTION)

49

D-29 = WAYFARER
(115" WB)

D-30 = MEADOWBROOK
and CORONET
(123½" WB)

(SAME WBs
THROUGH
'52)

TOTALLY RESTYLED
(INTRO. 2~49)

$2865.

CORONET WAGON

new
103 HP @
3600 RPM
(TO '53)

new
SWITCH KEY STARTING

LONGER on the inside . . . SHORTER outside!
WIDER on the inside . . . NARROWER outside!
HIGHER on the inside . . . LOWER outside!

150

DODGE

new DIPLOMAT H/T

WAYFARER ROADSTER
$**1727.**

$**2233.**

"GYRO-MATIC" AUTO. TRANS. AVAILABLE ON CORONET MODELS (OPTIONAL)

INTERIOR

BACK SEAT

SUPER-SIZE LUGGAGE COMPARTMENT!

4-DR. SEDAN

REAR WINDOW ENLARGED

MEADOWBROOK $1848.
CORONET $1927.

50   D-33 = WAYFARER
D-34 = MEADOWBROOK ; CORONET

(2 TONE COLORS AVAIL.)

PRODUCTION: WAYFARER
BUSINESS CPE. (7500); 2-DR. SED. (54,597);
SPORTABOUT ROADSTER (2903)
MEADOWBROOK 4 DR. SEDAN } (221,791)
CORONET 4 DOOR SEDAN }
CLUB CPE. (38,502); CONVT. (1800);
DIPLOMAT H/T (3600); WAGON (WD.) (600);
WAGON (STEEL) (100) (KNOWN AS "SIERRA")

new GRILLE with FEWER and HEAVIER PIECES

(8-PASS. SEDAN AVAIL. ON 137½" W.B.)

151

DODGE

PRODUCTION, 1951-1952 : WAYFARER = SPORTABOUT RDS. ('51 ONLY) (1002);
BUSINESS COUPE (6702); 2-DR. (70,700);
MEADOWBROOK, CORONET SEDANS (COMBINED, 329,202);
CORONET = CLUB COUPE (56,103); CONVERTIBLE (5550); DIPLOMAT H/T (21,600);
SIERRA WAGON (4000); 8-PASSENGER SEDAN, 137½" W.B. (1150)

51

WAYFARER

$1936.

DASH

ON 1951 MODEL, ENTIRE GRILLE IS CHROMED (INCL. LOWER PIECE)

FEATHER-TOUCH BRAKING!
Big Safe-Guard Hydraulic Brakes stop smoothly, surely, safely. Cyclebond linings, with their larger braking surface, last up to twice as long. New feather-touch parking brake holds securely on even steep grades . . . easily released with a twist of the wrist.

D-41 = WAYFARER

D-42   MEADOWBRK.
CORONET

$2059. UP

SHOWN IN SAN FRANCISCO, LOOKING EAST TOWARD OAKLAND.

$2256.

CORONET

D-41 and D-42 SERIES CONTINUE with LITTLE CHANGE

CORONET SIERRA WAGON $2908.

new HUBCAPS/WHEEL CVRS.

1952 SERIAL NOS. START AT:
WAYFARER   MEAD./CORONET

37175001 (Detroit)   31867601
48009901 (San Leandro) 45090601
48507601 (Los Angeles)
45527501 =
MD., COR., (L.A.)

LOWER PART OF GRILLE IS PAINTED.

52

CORONET "DIPLOMAT" H/T $2602.

FINAL YEAR FOR WAYFARER MODEL; REPLACED IN '53 BY MEADOWBROOK SPECIAL

DODGE

MEADOWBROOK 6
D-46

2000.

6 CYL.
MEADOWBROOK SEDAN

D-47

(15,751 BLT.)

2-DR.
SUBURBAN

$2176.

CVT.
2494.

D-48 (114"WB)
D-44 (119")
CORONET V8

new 241.3 CID
V8 (THROUGH '54)

140 HP
@ 4400
RPM

('00 BLT.)

Sensational New
140 Horsepower RED RAM V-8 ENGINE!

(124,059 BLT.)
$2220.

CORONET SEDAN

6
OR V8
53
(RESTYLED)

new V8

INTERIOR →

$2198.

CORONET V8
DIPLOMAT
HARDTOP
(7,334 BLT.)

2361.

2 VIEWS

WIRE WHEELS,
CONTINENTAL SPARE
AVAIL.

CLUB
CPE.
(32,439
BLT.)

ABOUT 56%
OF 1953 DODGES
SOLD WERE
V8s.

V8 MODELS
have

DODGE V EIGHT

BELOW
RAM HOOD ORNAMENT

114" OR 119" WB
(THROUGH '54)

153

DODGE

SUBURBAN
2-DOOR
WAGON
(CORNT.
V8)
(3100 BLT.)
$2517.

$2136.

CORONET 6

$2109.

D-51, D-52 (6 CYL.)

6-CYL. NOW HAS
110 HP @ 3600 RPM

54

D-50,
D-53 (V8)

$2349.
(8900 BLT.)

CLUB
CPE.

ROYAL
SERIES
IS new !

new GRILLE   ROYAL V8

ROYAL 500 CVT. IS PACE CAR AT 1954 INDY 500 RACE.

H/T

ROYAL V8
SEDAN

2373.
(50,050 BLT.)

V8 has 140
OR 150 HP @ 4400 RPM
(7.1 OR 7.5 COMPR.)

VARIOUS INTERIORS
(JACQUARD FABRICS)

DEPENDABLE
NEW '54 DODGE
Elegance in Action

Fully-automatic PowerFlite and full-time
Power Steering—yours at moderate extra cost.

PRODUCTION : MEADOWBROOK 6 = CLUB CPE. (3501) ; 4-DR. SEDAN (7894)
          MEADOWBROOK V8 = " " (750) ; " " (3299)
CORONET 6 = CLUB CPE. (4501) ; 4-DR. SEDAN (14,900) ; SUB. WAGON (6389) ; SIERRA 4-DR. WAGON (312) (6 OR 9 PASS.)
CORONET V8 = " " (7998) ; " " (36,063) ; " " (3100) ; " " " (988) " " " "
SPT. H/T (100) ; CVT. (50)   ROYAL V8 = CLUB CPE. (8900) ; 4-DR. SEDAN (50,050)
" " (3852) ; " (1299) "500" CVT. (701)

154

DODGE FLASHES AHEAD IN '55

SEDAN CORONET

(6- 15,976 BLT.)
(V8 - 30,098 ")

REAR

$2452. (4867 BLT.)
CORONET V-8 2-DOOR SUBURBAN

$2761. (5506 BLT., 6 or 8 PASS.)
ROYAL V-8 4-DOOR 8-PASSENGER SIERRA

V8 ENGINE NOW 270 CID

$2473. CUSTOM ROYAL V-8 4-DOOR SEDAN

2-TONE PAINT JOBS AVAIL.

new CUSTOM ROYAL LANCER H/T (30,499 BLT.) H/T

$2543.

D56-1= CORONET 6
6 CYL., 230.2 CID
123 HP @ 3600 RPM

D55-1= CORONET V8
and
D55-2= ROYAL V8
270.1 CID V8
(175 HP @ 4400 RPM)

55

TOTALLY RESTYLED

(MID-1955 "LA FEMME" MODEL FOR WOMEN, "TEXAN" MODEL FOR MEN.)

D55-3= CUSTOM ROYAL V8
270.1 CID V8 (183 or 193 HP @ 4400 RPM)

2-TONE COLORS on DASH; A/T CONTROL on DASH

$2516.

CUSTOM ROYAL LANCER SEDAN (new)

WITH THE FORWARD LOOK

new 120" WB (ON ALL, THROUGH '56)

6 CYL. NOW HAS 131 HP @ 3800 RPM (230 CID)
V8 HAS 189 to 340 HP (270, 315 or 354 CID)

D-62 (6 CYL.)
D-63 (V8)

$2807.

56

new FINS and EMBLEM in GRILLE
new SIDE TRIM
DIPS at REAR
new BUMPERS
HIGHER REAR FENDERS

In all the world no car like this
The New Dodge Lancer goes 4 door!

CUSTOM ROYAL LANCER 4 DR. H/T

The look, the feel, the power of success: New '56 Dodge Custom Royal Lancer 4-Door

MORE 1956 DODGES ON NEXT PAGE

New '56 DODGE
(CONT'D.)
VALUE LEADER OF THE FORW?

SIERRA $2438.

CORONET LANCER V8

$2716. UP

ROYAL

$2194. CORONET 6

$2513.

$2623.

DASH

CUSTOM ROYAL SEDAN

8-PASS. CUSTOM SIERRA

CUSTOM ROYAL   CONVERTIBLE $2913.

"LA FEMME" WOMEN'S EDITION CONT'D. FR. MID-1955 (1100 BLT., PRICE $2993.)

WITH 2-TONE ORCHID PAINT

PUSHBUTTON POWERFLITE, greatest advance in driving ease and control. Proven by years of successful testing!

Dodge

new TYPES OF PUSH-BUTTON TRANSMISSION CONTROL

2. DR. H/T CUSTOM ROYAL LANCER

$2693.

REAR CLOSE-UP

MODEL SERIES				
D 62 CORONET 6	(142, 613 BLT.)	$2194. to $2491.	230.2 cid 6 (131 HP)
D 63-1 CORONET V8	(IN ABOVE FIGURE)	$2302. to $2822.	270 cid V8 (189 HP)
D 63-2 ROYAL V8	(48,780 BLT.)	$2513. to $2974.	new 315 cid V8 (218 HP)
D 63-3 CUSTOM ROYAL V8	(42,293)	$2623. to $2913.	(SAME)
(also 260 HP, 315 cid V8 OPTIONAL IN ANY MODEL.)			

57 Dodge $2370.

SWEPT·WING $2861.

2-DR. SUBURBAN (D-70)

(TOTALLY RESTYLED)
D-72 (6)   D-66, 67, 70 (V8)

new 325 or 354 cid V8s

4-DR. SIERRA (D-70) (D-71 IS CUSTOM SIERRA)

CORONET (6 or V8)
(D-72 or D-66)

$2818.

ROYAL LANCER (D-67-1) 2-DR. H/T

$2991. CUSTOM ROYAL LANCER (D-67-2)

new 7.50 × 14 TIRES (8.00 × 14 WAGON, CVT.)

38 to 40 HP

new 122" WB (ON ALL MODELS, THROUGH '59)

new COMPOUND-CURVED WINDSHIELD

ADDED LOWER "TEETH" IN GRILLE OF ABOVE LATER MODEL.

REAR FINS "OVERLAP" FENDER

138 TO 333 HP

325, 350 or 361 CID V8s

CORONET

CUSTOM ROYAL LANCER

ROYAL

$2797.

$3071.

LD-1 (6)
LD-2, LD-3 (V8)

58 new GRILLE

EARLY MODEL '58s SHOWN ABOVE

SIERRA

PRODUCTION =
CORONET 6, V8 (77,388) $2449.-$2942.
ROYAL V8 (15,165) $2797.-$2915.
CUSTOM ROYAL V8 (25,000+) $3030.-$3298.
STATION WAGONS, V8 (20,196) $2970.-$3354.

SPRING SWEPT·WING

by **Dodge** ← 58½ MODEL with IDENTIFYING CHARACTERISTICS

PAINTED HEADLIGHT AREA, also new GRILLE MEDALLION ON 58½

new colors

157

'59 DODGE

CORONET 6, V8 #21 2 DR. $2516. (6); $2636. (V8)

59½ Silver Challenger two-door sedan

FINAL L-HEAD 230 CID 6 REDUCED TO 135 HP @ 3600 RPM

6 OR V8

SIERRA

326, 361 OR 383 CID V8s have 255 HP @ 4400 RPM TO 345 HP @ 5000 RPM

#43 LANCER 4-DR H/T

6-PASS. $3103. #45-A
9-PASS $3224. #45-B

#27 CONVERTIBLE

CUSTOM ROYAL $3270.

$3422.

new TAIL-LIGHTS

SWIVEL-SEATS AVAIL. (new)

INTERIOR (CUSTOM ROYAL CVT.)

59

MD1-L (6)

MD2-L, MD3-M, MD3-L, MD3-H (V8)

PUSH-BUTTON SHIFT PLAN ←

Custom Royal

DASH

new interior

PRODUCTION:
MD1-L CORONET 6, MD2-L CORONET V8 (96,782) $2516.-$3089.
MD3-M ROYAL V8 (14,807) $2934.-$3069.
MD3-H CUSTOM ROYAL V8 (21,206) $3145.-$3422.
MD3-L SIERRA V8 WAGONS, MD3-N CUSTOM V8 WAGONS (23,590) $3103.-$3439.

'60 DODGE

#45-A, B MATADOR WAGON

$3239. UP

MATAD. -DR H/T 996.

$3075.

#43 MATADOR 4-DR. H/T

122" WB ON LARGE DODGES (THROUGH '61)

MATADOR·POLARA

$3506. UP

361 OR 383 CID V8s (THR.'61) IN ALL DODGES EXCEPT new DART OR '61-2 LANCER

(295 OR 330 HP)

#27 CVT.

NEW H.V. SLANTED 6-CYL. ENGINE AVAIL. ONLY new SUBSIDIARY DART.

#43 POLARA 4-DR. HARDTOP

$3416.

POLARA has BRIGHTWORK ON LOWER REAR ← FENDER.

3275.

PD1 and PD2 SERIES

new DART LISTED SEPARATELY.

60 RESTYLED

POLARA has HEAVY BAND ATOP FENDER ↓

UNIBODY CONSTRUCTION

SEE ALSO **DART.**

TOTALLY new FRONTAL APPEARANCE, WITH V-DIP IN LOWER EDGE OF GRILLE.

PRODUCTION :
PD1-L MATADOR	(27,908)	$2930. —	$3354.
PD2-H POLARA	(16,278)	$3141. —	$3621.
(CONVERTIBLE AVAIL. ONLY IN POLARA SERIES) = $3416.			

DODGE

265 TO 330 HP

122" WB

POLARA CONVERTIBLE V8

DASH

POLARA 2 DOOR HARDTOP V8

61

RESTYLED

14,032 BLT.

POLARA HARDTOP WAGON V8,
6 OR 9 PASSENGER

POLARA V8
IS
ONLY LARGE
SERIES

SEE ALSO DART,
OR
LANCER

MODELS, PRICES #542 H/T $3032.
#543 4 DR. SEDAN $2966.
#544 4 DR. H/T $3110.
#545 CONVERTIBLE $3252.
#578 6-PASS. WAGON $3294.
#579 9-PASS. WAGON $3409.

$3019. #542 2-DR. H/T

Dodge Polara 500—2 dr Hardtop

$2960. #544 4-DR. H/T

Dodge Polara 500—4-dr Hardtop

POLARA 500

POLARA 500

361 OR
413 CID
V8
305
OR
380
HP

(SD2-P)
POLARA MODELS
TOTALLY RESTYLED
116" WB (AS ON
DART)

62

$3268.

#545 CVT.

#614 CUSTOM 880
4-DR. H/T

$3109.

#658 6-PASS. CUSTOM 880 WAGON
#659 9-PASS.

CUSTOM 880

(CONSERVATIVE
OLDER TYPE
STYLING) 122" WB

(SD3-L)

361 CID V8
265 HP @
4400 RPM

$3292. OR $3407.
6 PASS.   9 PASS.

PRODUCTION:
SD2-P POLARA 500
(12,268) $2960.-$3268.
SD3-L CUSTOM 880
(17,505) $2964.-$3407.

$3030.

#612 CUSTOM 880 2-DR. H/T

160

1963 DODGE

330 SERIES

$2648. UP

330 2-DR. SED.
$2245. UP

$2470. UP

$2854. UP

440 WAGON

440 SERIES

$2381. UP

116" OR 119" WB

$2438. UP

225 CID 6 (145 HP @ 4000)

318, 383 OR 426 CID V8
(230 TO 425 HP)

DASH
$3196.

(V8) 4-DR. H/T
$2781.

POLARA SERIES

POLARA 500 SERIES
BUCKET SEAT

#642 H/T
$2965.

TD SERIES
63

#L-558, 559
$3292. UP

#L-514 4-DR. H/T

CUSTOM 880

361 OR
383 CID V8
(265 OR 305 HP)

122" WB

$3109.

REAR VIEW OF CUSTOM 880 WAGON

PRODUCTION:
TD1-L 330 (6)	(51,761)	
TD2-L 330 (V8)	(33,602)	
TD1-M 440 (6)	(13,146)	
TD2-M 440 (V8)	(49,591)	
TD1-H POLARA (6)	(68,262)	
TD2-H POLARA (V8)	(40,323)	
TD2-P POLARA 500 (V8)	(7256)	
TA3 880 (V8)	(9831)	
(CUSTOM 880) (V8)	(18,435)	

'64 Dodge

225 CID 6 OR 318, 383 OR 426 CID V8

440 (VD-2)

7.00 × 14 TIRES

$2264. UP

330

$2637. UP

POLARA

$2994.

SPORTSMAN WAGON

145 TO 425 HP

PRODUCTION:
VD1-L 330 (6) (57,957)
VD2-L 330 (V8) (46,438)
VD1-M 440 (6) (15,147)
VD2-M 440 (V8) (68,861)

VD1-H POLARA (6) (3810)
VD2-H POLARA (V8) (66,988)
VD2-P POLARA 500 (V8) (17,787)
VA3 880 (V8) (10,526)
(CUSTOM 880 (V8) (21,234)

POLARA   A- PARK. LOCK ; B- TRANSMISSION BUTTONS ; C- SPEEDO. ; D- CLOCK ;
E- HEATER CONTROLS ; G- RADIO ; H- GLOVE BOX ; I- ASHTRAY, LIGHTER ;
J- IGNITION ; K- WIPER CONTROL ; L- LIGHTS ; M- PARK. BRAKE RELEASE

DASH

note THAT POLARA DASH (ABOVE) DIFFERS
FROM 880 DASH (SEE 880 CVT., BELOW)

$3155. UP

880 WAGON

$2826.

4-DR. SED.

CONCAVE GRILLE ON 880

$3264.

WRAPAROUND TAIL LIGHTS ON 880

(VA-3)

8.00 × 14 TIRES

880

64

FINAL YR. OF '61-STYLE ROOFLINE ON 880

Dodge AW1 (6) AW2 (V8) **Coronet**

Dodge Coronet 440 2-door hardtop. 6 and V8 power.

16" and 17" WB

$2622. UP

Coronet

CORONET DASH

Dodge Coronet 440 Station Wagon (3-seat model also offered)

7.75 × 14 TIRES

7.35 × 14 TIRES

$2403. UP

$2674.

$2913. **Polara** #D-14 (AD2-L)
4-door hardtop. V8 power.

145, 180, 230, 265, 270, 315, 330
340, 365 OR 425 HP.

V-8 CHOICES INCLUDE 273, 318, 361, 383, 413 OR 426 CID

new 121" WB

Monaco two-door hardtop.

Dodge Monaco. Limited edition. 2-door hardtop. V8 power.

(13,096 BLT.)

$3355.

MONACO DASH

65 Monaco
(new)(H/T only) (AD2-P)

8.25 × 14 TIRES

CUSTOM 880
4-DR. SED.

$3010.

CUSTOM 880

4 DR. H/T

$3150.

CUSTOM 880 (AD2-H)

Wagon has 8.50 × 14 TIRES

$3422.

44,496 CUSTOM 880s BLT. (6 MODELS)

PRODUCTION (OTHER MODELS):
AW1-L CORONET (6) $2217. UP
AW2-L CORONET (V8) $2313. "
AW2-N CORONET 440 (6) $2377. UP
" CORONET 440 (V8) $2473. UP

AW2-P CORONET 500 (V8) (32745) $2674. UP
AW2-L POLARA (V8) (12,705) $2730. UP

(WEST COAST)
$2736.

$2264. UP

DODGE

STD. CORONET SERIES (lowest priced Coronets
(NO SIDE CHROME)

2-DR.

$2303. UP

$2722.

4 DR. WAGON

CORONET 440

Coronet DELUXE
(2 DR. WAGONS ALSO)

ENGINES:
225 CID 6 (145 HP); 273½ CID V8 (180 HP);
318 CID V8 (230 HP);
361 CID V8 (265 HP);
383 CID V8 (270 OR 325 HP);
440 CID V8 (350 HP)

$2264. ~ 3604.
PRICE RANGE
TOTAL 1966 PROD.:
419,287
CALENDAR YR. = 435,026 *
* INCL. DART, CHARGER

ALL CORONET MODELS WITH
225 CID 6,
273½ CID V8
318 CID V8
361 CID V8 OR
383 CID V8

JOIN THE DODGE REBELLION

66
(6 CYL. OR V8)

new GRILLES

Signal when ready. Hidden behind the handsome grille are Coronet's turn signals and parking lights.

2457. UP

117" WB   Coronet 440
$2891.

$2611.

CORONET 440

440 H/T INTERIOR (ABOVE)

← Coronet 500 →

$3045. (WEST)

Coronet 500.

$2827. UP
$3261. (WEST)

500 DASH

CORONET

(CONT'D. NEXT PAGE)

DODGE

3/83. UP

POLARA V8

Polara V8 CVT. 3161.

66 (CONT'D.)

FROM $3555. (WEST)

$2948.

4-DR. H/T

$3320. (WEST)

$3533. (WEST)

Polara V8

V8 ("POLARA 500" MODELS ALSO)

121" WB and V8 ENGINES=IN EACH POLARA, MONACO)

(WEST) $3405.

$3033.

Slip gracefully and frugally out of the low-price field. It's never been so easy. See what Dodge Monaco and Polara cars offer you as standard equipment. Then slip behind the wheel—and get a kick out of driving!

Monaco 4 DR. SEDAN

2 SEAT or 3 SEAT 4 DR. $3436. OR 3539.

MONACO V8

FROM $3808 (WEST)

wagons

A-60

REAR DECK BEARS "MONACO" NAME, INSTEAD OF "DODGE"

(WEST) $3759.

$2884.

90" WB

Monaco 500 2-door hardtop. 383 4-barrel V8 power, standard.

$3604.

MONACO 500—THE VERY FINEST DODGE OF ALL FOR '66.

CUSTOM SPORTSMAN

($2567. STD. SPORTSMAN ALSO)

MODELS: CORONET BWL (66,161 BLT.) (INCLUDES CORONET DELUXE); CORONET 440 BWH (128,998 BLT.); CORONET 500 BWP (55,683 BLT.); POLARA BD2-L (107,832 BLT.); MONACO BD2-H (49,773 BLT.) MONACO 500 HARDTOP (ONLY BODY TYPE IN MON. 500 BD2-P SERIES)(10,840 BLT.)

SMALL VANS LIKE CUST. SPTSMN. (ILLUSTRATED ABOVE) FOUND FAVOR DUE TO THEIR ROOMY INTERIOR SPACE, SHORT EXTERIORS.

DODGE _Coronet_

$3070.

Dodge Coronet 2-seat station wagon.

$2622. UP

$2693. UP

Dodge Coronet Deluxe 2-seat station wagon.

$3141.

440   **DLX.**

FRONT CLOSE-UP

$2397. UP
COR. DLX.

(INTRO. 9-29-66)

67
new GRILLES

440

$3352.

CORONET R/T

(w. 440 CID V8,

FROM $3666. [WEST]

COR. 500 TAIL-LTS. APPEAR INVISIBLE BY DAY.

Dodge Coronet 500 SE (Special Edition)

Coronet 500
The Dodge Rebellion wants you!

$3183. UP

Dodge Polara station wagon, available in 2-seat and 3-seat models.

POLARA STATION WAGON

POLARA REAR LIKE MONACO

MONACO

$3170. UP

(POL. 500 _has_ VERTICAL TRIM STRIPS NEAR FRONT TIP OF FRONT FENDER.)

$3676. [WEST]

Monaco 500
BEARS "DODGE" NAME AT REAR

"MONACO" NAME AT REAR, (EXCEPT ON MONACO 500)

$3896. [WEST]

CORRUGATED STRIP ALONG SIDE
(WEST COAST PRICES IN SMALL PRINT)

$3712.

$2359. ~ 3712.
PRICE RANGE

361 CID V8 DISCONTINUED

TOTAL 1967 PRODUCTION: 295,449

DODGE $2487. TOTAL 1968 PRODUCTION: 455,761 $3379.

$2525. CORONET DELUXE

PLAIN TOP CORONET R/T

$3212. UP

a two-way tailgate... CORONET 440

$2603. VINYL TOP $3613 440

WAGON $2924. UP

Coronet 500 wagon

FROM $3933. (WEST)

new GRILLES

H/T $2879. CORONET 500

(INTRO. 9-14-67) Watch out. You're getting **68 Dodge fever**

POLARA 122" WB (SINCE '67)

$3483

POLARA WAGON FROM $4105

POLARA 318 CID V8 (230 HP)

MONACO 500 H/T (4568 BLT.)

MONACO WAGONS FROM $4469.

383 CID V8 (290 HP)

new SIDE SAFETY LTS. (CIRCULAR)

$4035.

new FULL-WIDTH TAIL-LT. ASSEMBLY ON POLARA, MONACO.

$3432. 4 DR. H/T $3815. (WEST)

CUSTOM **SPORTSMAN**

108" WB (AVAIL. SINCE '67)

two sizes— 4 models— to choose from!

(90" WB $2887. UP)

MONACO PRICE RANGE $2487.~3869.

167

DODGE
CORONET

$2692.

440 H/T $3521. (WEST)

new GRILLES

REAR DETAIL

69 (INTRO. 9-19-68)

new TAIL-LIGHTS

Coronet. 500 V8

(WEST) $3655.

Coronet 500

FRONT VIEW

White Hat Special Coronet.

The Dodge Coronet White Hat Special comes in a 2-door hardtop or 4-door sedan—with the features listed below—at a special low package price.
- Vinyl roof in black, white, tan, green—or standard top
- Whitewall tires ■ Front, rear bumper guards ■ Deep-dish wheel covers ■ Light group ■ Outside, remote-control rearview mirror ■ Bright trim package.

H/T
$2929.

$3138. (SUPER BEE, BELOW)

note DUAL HOOD SCOOPS on

(32,050 CORONET 500s BLT. '69) (H/T, CVT., SED., WAGON)

TOTAL 1969 PROD.: 611,645

'69 CORONET SUPER BEE ↓

STANDARD SUPER BEE EQUIPMENT
- Special 4-bbl. 383-cid Magnum V8 (440 Magnum V8 heads, valve gear, hot cam and manifolds), 335 hp @ 5,200 rpm • Dual exhaust
- Hurst 4-speed with HD clutch • HD suspension
- HD shocks • HD brakes • Dodge Charger Rallye instrument panel

OPTIONAL
- 426 Hemi—two 4-bbl. carbs—425 hp @ 5,000 rpm

REAR AXLE RATIOS
- 383 Magnum V8—standard: 3.23:1; optional: 3.55:1, 3.91:1
- Hemi—standard: 3.23:1; optional: 3.54 (with 4-speed manual), 3.55:1 (with automatic), 4.10:1 (with manual or automatic)

H/T $3697. (WEST) (note BEE FIGURE ON GRILLE)

THIS DECAL and POPULAR ADVERTISING FIGURE IS WELL KNOWN, BUT DIFFERS FROM BEE ON GRILLE.

(27,846 BLT.)

(COUPE or H/T)

EARLY "SUPER BEE" →

THIS TYPE AVAIL. in LATE '68 MODEL YEAR.
383 CID V8 (335 HP @ 5200) OR
426 CID HEMI V8 (425 " @ 5000)

(CONT'D. NEXT PAGE)

DODGE

$3188.

4 DR. H/T

Polara.

a 230-hp V8. Not to mention an all-new instrument panel and concealed windshield wipers.

DODGE WHITE HAT SPECIAL POLARA.

The Dodge Polara White Hat Special comes in a 2-door or 4-door hardtop—with the features listed below—at a special low package price.
• Vinyl roof in black, white, tan, green or standard top • Front, rear bumper guards • Fender-mounted turn signals • Outside, remote-control rearview mirror • Whitewall tires • Deep-dish wheel covers • Bright trim package

WHITE HAT SPECIAL

69
(CONT'D.)

$2554.~4046. PRICE RANGE

ENGS.:
225 cid 6 (145 HP)
318 cid V8 (230 HP)
383 cid V8 (290, 330 or 335 HP)
426 cid V8 (425 HP)
440 cid V8 (350 or 375 HP)

POLARA $3117.

500
3314.

TOTAL 1969 PROD.: 324,256

FINAL YEAR FOR POLARA 500

1969 POLARA

This year, **DODGE** is turning up the *fever*

Monaco.

MONACO GRILLE (left) DIFFERS ONLY SLIGHTLY FROM POLARA's.

$3917. UP

wagon

Dodge CHRYSLER MOTORS CORPORATION

In a test of acceleration, economy, and braking ability, a 1969 Monaco was overall winner, Class II, in the Union/Pure Oil Performance Trials.

INTERIOR

MONACO 4 DR. H/T

FROM $4707. (WEST)

"38,566 MONACOS BLT., INCL. 2-DR. H/T, SED.)

$4381. (WEST)

all-new aircraft-type instrument panel. And ahead of it all—a big 383-cu.-in. V8. 1969 Monaco.

$3591.

(1966-1978) Dodge CHARGER

66

$3469.

HDLTS. DISAPPEAR INTO GRILLE

BUCKET SEATS, FRONT and BACK

...new leader of the Dodge Rebellion.

117" WB (LIKE Coronet) (THROUGH '70)

V8s: 318 CID (230 HP) 361 (265 HP) 383 (325 HP) 426 CID Hemi (425 HP)

7.35 x 14 TIRES

(37,344 BLT.)

FASTBACK STYLING (THROUGH '67)

V8s: 318 CID (230 HP); 383 CID (325 HP); 426 CID Hemi (425 HP) new 440 CID Magnum V8 AVAIL. (375 HP)

$3128. ($3482., WEST COAST)

VINYL TOP OPT.

67

XP-29 MODEL CONTINUES WITH FEW CHANGES

the Dodge Rebellion.

15,788 BLT.

Join the fun . . . catch
DODGE *fever*

DASH

Dodge Charger

$3184. UP

68

(RESTYLED)
NO LONGER
A FASTBACK

new
SIDE
SAFETY LTS.
(ROUND)

FROM
$3371.

3592.

0,057
BLT.)

R/T SE

STANDARD CHARGER R/T EQUIPMENT
- 440-cid Magnum (4-bbl.) V8, 375 hp
- Choice of 3-speed automatic or Hurst
  4-speed manual • Dual exhausts
- HD suspension • HD shocks • HD brakes
- Dodge Charger Rallye instrument
  panel • F70x14 wide-treads

OPTIONAL
- 426 Hemi

69

new SPLIT
GRILLE

Success Car of the Year

117" WB

$4000.

UNIQUE REAR
"TAIL"
SPOILER

Charger Daytona:

DAYTONA

RARE!
(505 BLT.)   **Totally new**

69½-70

(1970 PLYMOUTH ROAD-RUNNER
"**SUPERBIRD**" SIMILAR)

(TOTAL 1969 PROD. [ALL MODELS]
AMOUNTS TO 90,000-PLUS.)

EDSEL $2519. UP

MFD. BY FORD MOTOR co
(1958, 1959, 1960 MODELS ONLY)

ROUNDUP #26
2-DR. WAGON
$2876.

WAGONS HAVE SPECIAL TAIL-LIGHTS

note "HORSECOLLAR" CENTER GRILLE

SLOGAN: "THIS IS THE EDSEL"
(OTHER SLOGANS ALSO)

"Teletouch AUTO. TRANS CONTROL BUTTONS in STEERING-WHEEL HUB. (OPTIONAL) (1958 ONLY)

58

DASH → (with REVOLVING SPEEDOMETER)

#47, 48

#27, 28
VILLAGER 4-DR. WAGON
$2933. UP

V-8s
361 CID, 303 HP
OR 410 CID, 345 HP

BERMUDA 4-DR. (DELUXE WAGON with WOODGRAIN)
$3190. U

#84 4 DR. H/T
$3615.

$380/

RANGER (LOWEST-PRICED)

#21 2-DR.
(has MINIMUM of SIDE CHROME)
$2519.

CITATION

PACER
#42 4 DR.

$2735.

$3346.

#63 H/T
CORSAIR

#85 CONVT.

CITATION
(TOP of LINE) has INSET PANEL SET WITHIN REAR FENDER TRIM LOOP

W.B.s
116" (WAGONS)

118" (RANGER PACER)

124" (CORSAIR CITATION)

PRODUCTION: RANGER = 2 DR. (4615); 4 DR. (6576); H/T (5546); 4 DR. H/T (3077); ROUNDUP WAG. (963);
VILLAGER, 6 PASS. (2294); VILLAGER, 9 PASS. (978) PACER = 4 DR. (6083); H/T (6139); 4 DR. H/T (4959);
CONVT. (1876); BERMUDA, 6 PASS. (1456); BERMUDA, 9 PASS. (779)
CORSAIR = H/T (3312); 4 DR. H/T (5880)
CITATION = H/T (2535); 4 DR. H/T (5112); CONVT. (930)

EDSEL

RANGER

PACER, CITATION, ROUNDUP and BERMUDA MODELS NO LONGER AVAIL.

$2629. UP

223 CID 6-CYL. OHV ENGINE ALSO AVAIL. STARTING 1959. (145 HP @ 4200 RPM)

3 V8 ENGINES:
Ranger 292 CID (200 HP @ 4500 RPM)
Express 332 CID (225 HP @ 4600)
or Super Express 361 CID (303 HP @ 4600)

CORSAIR

new ROUND TAIL-LIGHTS

VILLAGER

new DASH

CORSAIR has VERTICAL CHROME STRIPS

59 (RESTYLED)

SLOGAN: "MAKES HISTORY BY MAKING SENSE"

INSET ANODIZED TRIM PANEL IDENTIFIES 1959 CORSAIR.

new 120" W.B. (ALL MODELS, THROUGH '60)

new GRILLE INCORPORATES HEADLIGHTS, with HORIZONTAL MEMBERS IN CENTER SECTION.

RANGER = #57F 4 DR. H/T (2352); #58D 4-DR. (12,814); #63F H/T (5474); 2-DR. (#64-C) (7778)
CORSAIR = #57B 4 DR. H/T (1694); #58B 4-DR. (3301); #63B H/T (2315); #76E (CONVT.)
VILLAGER WAGONS = #71E 9 PASS. (2133); #71F 6 PASS. (5687)

EDSEL

SLOGAN : "NEW! NIFTY! THRIFTY!"

CORSAIR NO LONGER AVAIL.

RANGER and VILLAGER ARE ONLY MODELS OFFERED FOR EDSEL'S BRIEF 1960 SEASON.

#58 A
4-DR.
$2697.

#64 A
2 - DR.

EDSEL RANGER 2-DOOR SEDAN

new TAIL-LIGHTS PICK UP VERTICAL OVAL THEME FORMERLY DISPLAYED IN EDSEL GRILLE.

1960 MODEL PROD.:
WAGON, 6-PASS. (216)
WAGON, 9-PASS. (59)
2-DR. SEDAN (777)
4-DR. SEDAN (1,288)
2-DR. HARDTOP (295)
4-DR. HARDTOP (135)
CONVERTIBLE (76)
  TOTAL = ONLY
2,846
1960 EDSELS
BUILT!

2-DR. IS LOWEST-PRICED 1960 EDSEL. **$2643**. (OR $2635.30)

DASH

RESTYLED
60

$2989. UP

RANGER

#63A H/T
$2705.

new SPLIT GRILLE

#71 F (6 PASS.)
71 E (9 PASS.)
VILLAGER

new DASH

3 ENGINE CHOICES :
Economy 6 223 CID
  (145 HP @ 4000 RPM)

Ranger V8 292 CID
  (185 HP @ 4200 RPM)

Super Express V8 352 CID
  (300 HP @ 4600 RPM)

76-B
CVT.

$3000.

1960 EDSELS DISCONTINUED NOV., 1959

FAIRLANE
BY
(FORD)

new COMPACT / INTERMEDIATE **6** OR **V-8** FOR 1962 ; FORMERLY A FULL-SIZED FORD SERIES

2-DR.
2154.

"500" 2-DR.   **$2242.**

115½" w.b. (THROUGH '65)

"500" 4-DR.

$2304.

62 new

223 CID 6 (138 HP)
2.21 CID V8 (145 HP)
2.60 CID V8 (164 HP)

$2392. UP
(WEST COAST)

EASTERN PRICES FROM $2154.

PRODUCTION : 54-A 4 DR. (45,342); 62-A 2-DR. (34,264) ; 54-B "500" 4 DR. (129,258); 62-B "500 2-DR. (68,624); 62-C "500" SPT. CPE. (19,628)

SERIES 30 FAIRLANE (INCLUDES WAGONS)
$2154. - $2781.

SERIES 40 FAIRLANE "500"
$2242. $2504.

$2304.

500 new CONCAVE GRILLE

$2525. UP
SQUIRE →

RANCH WAGON

63

new 200 CID 6 (116 HP)
2.21 CID V8 and 260 CID V8 CONT'D.

$2324.

63½ AVAIL. WITH new VINYL ROOF

$2781.

new 289 CID V8 (271 HP)

H/T INTERIOR

175

FAIRLANE

"289" V-8 option

Fairlane wagons:

500 H/T

$2341.
$2502.

64 new GRILLE

CUSTOM RANCH WAGON (24,962 BLT.) $2612.

SEE ALSO:

FORD

PRODUCTION:
SERIES 30 = 2 DR. (20,421); 4 DR. (36,693); RANCH WAGON (20,980)
" 40 "500"= 2 DR. (23,447); 4 DR. (86,919); H/T (42,733);
H/T (BUCKET SEATS) (21,431); CUSTOM RANCH WAGON (24,962)

LOWEST-PRICED SERIES 30 2-DR. $2230.

| 1965 FAIRLANE PRODUCTION : 251,647 |

optional features. 4-speed stick. Overdrive. Tachometer.

500 H/T (41,405 BLT.)
w. BUCKET SEATS (15,141)

500 WAGON (20,506 BLT.)
$2648.

HEADLIGHTS PAIRED IN PLATES

271 solid-lifter horsepower high-shift automatic!

SERIES 30 4-DR.
(25,378 BLT.)
$2271.

NO MORE
RAISED "AIR
SCOOP"
EFFECT
ON
HOOD

65 new GRILLE

500
2-DR.
(16,092 BLT.)
$2312.

new OBLONG TAIL-LIGHTS

176

Fairlane
$2533.
(23,942 BLT.)

new GRILLE, VERT.-STACKED HDLTS.

H/T 500-XL 116" WB GT

ENGINES:
200 CID 6 (120 HP)
289 CID V8 (200 HP)
OR 390 CID V8 IN
GT (335 HP)

66

1966

Fairlane GT –
FRONT END DETAIL

NAME ON REAR FENDER

$3068.
GT

FORD Fairlane

Special GT and GTA identification

GT **GTA**

new TAIL-LIGHTS

convertible (4327 BLT.)

(GTA has AUTO. TRANS.)

GT DASH

$2718. (INTRO. 9-30-66) FAIRLANE CLUB CPE. (10,628 BLT.) $2297.

67 116" WB

new TAIL-LIGHTS and GRILLE 500

WAGONS have 113" WB

Fairlane 500 Wagon

$3163.

FAIRLANE'S SQUIRE WAGON AVAIL. SINCE '66

$2377. CLUB CPE. (8473 BLT.) (1943 BLT.) CVT. $3289. (WEST)

$2950.

2902. (8348 BLT.)
SQUIRE $3347. (WEST)

15,902 BLT.) (WEST)

Fairlane 500/XL

390 BADGE INDICATES A 390 CID V8

TOTAL (1967 PROD.: 190,383)

(14,871 BLT.) H/T

500/XL

$3063. (WEST)

GT

(note SIDE PAINT STRIPES ON GT)
$2839.

INTERIOR (500/XL)

Fairlane 500/XL Interior

Fairlane / TORINO (new!)

FAIRLANE

$2464. FAIRLANE SEDAN

$2962. (WEST)

H/T $ 2909.
FORMAL H/T $3156

TORINO.

FAIRLANE 500

$301 (WEST)

(WESTERN PRICES ABOVE)

$2710.

68 (INTRO. 9~22~67)

$2747.

TORINO INT.

TORINO GT

302 CID V8 (210 HP) OR
390 CID V8

(74,135 BLT.)
FASTBACK

Torino Squire

FAIRLANE

$2488. ~ 3107. PRICE RANGE

$2733.

↑2488.

TORINO

REAR SIDE SAFETY LT'S. SMALLER, MOVED FORWARD AND DOWN.

(INTRO. 9-27-68)

69 new GRILLES

250 CID 6 (155 HP)
302 CID V8 (220 HP)
351 CID V8 (250 OR 290 HP)
390 OR 428 V8s ALSO (TO 360 HP)

SPORTS-ROOF (FASTBACK) (61,319 BLT.)

$2840.

TORINO

1969

TORINO GT

178

FALCON (compact) $1912. and up (1960 – 1970½ MODELS)

BY FORD

60

4 DR.

$1974.

2 DR.

$1912.

4 DR. WAGON $2287.

CHOICE OF 2-DR. OR 4-DR. WAGONS

109½" WB
6.00 x 13 TIRES

2-DR. WAGON $2225.

6 CYL. 144 CID OVERHEAD VALVE ENGINE 90 HP

Ford MOTOR COMPANY

PRODUCTION:
#64-A 2-DR. (193,470); 58-A 4-DR. (167,896); 59-A 2-DR. WAGON (27,552); #71-A 4-DR. WAGON (46,758)

1961 PRODUCTION: 4 DR. (159,761); 2-DR. (149,982); 2-DR. ECONOMY SEDAN (150,032); 2 DR. WAGON (32,045); 4 DR. WAGON (87,933) new FUTURA COUPE (44,470)
(FUTURA INTRO. SPRING, 1961)

$1976.

4-DR.

85 HP (STD.) $2227.

A CHOICE OF TWO SURGING "SIXES"!
STD. 144 CID OR new 170 CID 6
85 or 101 HP

2-DR.

$1914.

new FUTURA CONSOLE

(4-DR. WAGON ALSO AVAIL., $2270.)

FALCON TUDOR WAGON

FORD *Falcon* '61
WORLD'S MOST SUCCESSFUL NEW CAR
new CONVEX GRILLE

Futura
(1961½)
$2162.

FORD

1961

FUTURA has 3 DARTS on REAR FENDERS, and SPECIAL HUBCAPS.

$2273. **Falcon '62** BEST SHAPE ECONOMY'S EVER BEEN IN $2384. DELUXE 2-DR.

FUTURA

FALCON SPORTS FUTURA

FUTURAS HAVE SPECIAL FRONT FENDER TRIM, AS ILLUSTRATED.

4-DR.

$2133. DELUXE $2427.

(new) SQUIRE $2603.

62

STD. 2-DR. $2243.

(STANDARD MODELS HAVE NO SIDE CHROME TRIM.)

NEW CONVEX GRILLE has ALL-VERTICAL PIECES.

1962 PRODUCTION : 381,558

PRODUCTION :

54A 4-DR. (62,365); 62A 2-DR. (70,630); (THESE 2 STANDARD MODELS KNOWN AS "SERIES 0.")
FUTURA SERIES 10 : 62B 2-DR. (27,018); 54B 4-DR. (31,736); 63B H/T (28,496); 63C SPRINT H/T (10,479);
76A CVT. (31,192); 76B SPRINT CVT. (4602) SERIES 20 WAGONS : 59A 2-DR. (7322);
59B DLX. 2-DR. (4269); 71A 4-DR. (18,484); 71B DLX. 4-DR. (23,477); 71C SQUIRE (WOODGRAIN) (8269)

$1985. STD.

new 164 HP V8 ALSO AVAIL. (260 CID)

$2298.

STD. 2-DR.

63

new GRILLE

DE LUXE

$2384.

2-DR.

FUTURA

$2603. SQUIRE 4-DR.

$2116. 2-DR.
$2198. H/T

MID-1963
"SPRINT" V8
H/T and CVT. are NEW

$2470.

(new) CONVERTIBLE

180

FALCON

Lively new Sprint $2600. $2320.

THESE MODELS INTRODUCED IN MID-SEASON

63½

new scatback hardtop

$2622.

Squire
(6766 BLT.)

new wider tread

$2481. ↑ UP
$2325. ↙ UP

SPRINT NOW AVAIL. AS 6 OR V8.

64

(RESTYLED)

new '260' cu. in. V-8 power option

new longer springs

$2226. (FUTURA)
$2337. (SPRINT)

$2665. ↙

Squire

H/T

New battery-saving alternator.

65

1965 PRICE RANGE:
$2020. TO $2671.

13" or 14" WHEELS

new GRILLE

DASH

new 170 cu. in. standard Six with optional 3-speed Cruise-O-Matic transmission

---

PRODUCTION: 2-DR. (35,858); DLX. 2-DR. (13,824); 4-DR. (30,186); DLX. 4-DR. (13,850)
FUTURA - 2-DR. (11,570); H/T (25,754); 4-DR. (33,985); CVT. (8215); DLX. 4-DR. WAGON (12,548)
SPRINT - H/T (2806); CVT. (300)
WAGONS - (DLX. LISTED WITH FUTURA) 2-DR. (4891); 4-DR. (14,911); SQUIRE 4-DR. (6703)

Falcon

FUTURA CLUB CPE. (21,997 BLT.)

(1960 – 1970)

$2669. ($2781., FUTURA)

$2183.

FUTURA

new THINNER SIDE TRIM

$2328.

SPT. CPE. WITH VINYL TOP
$2555. (WEST COAST)

(20,289 BLT.)

66

new GRILLE

DASH

111" WB
170 CID 6 (105 HP)
200 CID 6 (120 HP) IN SPT. CPE., WAGON
289 CID V8 (180-200 HP)

6 CYL. OR V8 (SINCE '63)

(FALCON RANCHERO PICKUP ALSO AVAIL.)

CLUB CPE. $2060.

TOTAL 1966 PRODUCTION: 182,669

1966

$2284. (WEST COAST)
7.35/7.75 x 15 TIRES

TOTAL 1967 PRODUCTION: 64,335

289 CID V8 has 225 HP, OTHER ENGINES AS IN 1966.

new TIRE SIZE = 6.95/7.35 x 14"

$2437.

FUTURA SPTS. CPE. (7053 BLT.)

$2663. (WEST COAST)

2 new COWL INDENTATIONS ('67 ONLY)

67

(INTRO. 9-30-66)

new GRILLE

REAR (STD.)

182

FALCON

WAGON ('69)

$2301.

('68)

EARLY **68-70**

$2579.

170 CID 6 (100 HP)
289 CID V8 (195 HP)
200 CID 6 AVAIL. '69-
'70 (115-120 HP)
302 CID V8 (220 HP) '69-70

FALCON. 7 MODELS. MORE
THAN ANY OTHER COMPACT.

('70)

FUTURA SEDAN

111" WB
(WAGONS 113")

$2541.

FALCON
PRODUCTION
SUSPENDED
BECAUSE OF HIGH
COST OF ADDING
LOCKING STEERING
COLUMNS, IN
COMPLIANCE WITH
GOVT. SAFETY REGULATIONS.

INTERIOR ('69)

('69)

PRICE RANGE, 1968	$2252. ~ 2728.		PRODUCTION	
" " 1969	2283. ~ 2771.		1968 =	131,389
" " 1970	2390. ~ 2878.		1969 =	95,019
			1970 =	15,694

$2771. ('69)

FALCON TEMPORARILY REVIVED, AS A
BUDGET-PRICED MODEL OF
TORINO.

new E78/G78 x 14 TIRES

2-DR. $2460.

70½ NEW

LARGER 117"-WB
TORINO SERIES
(WAGON - 114")

3 FINAL FALCONS =
2-DR. $2827.
4-DR. $2867.
4-DR. WAGON $3163. (WEST COAST PRICES)

(DISCONTINUED SUMMER, 1970)

67,053 BLT.

STD. ENGINES:

new 250 CID 6
(155 HP)
302 CID V8
(220 HP)

AVAILABLE
351 CID V8 (250 or 300 HP)
429 CID V8 (360 or 370 HP)

26,071 2-DRS.; 30,443 4-DRS.; 10,539 4-DR. WAGONS

(SINCE 1967)

The Magnificent Five are here!

Firebird BY PONTIAC

$2667. UP

$3127. (WEST)

Firebird

STD. FIREBIRD 230 CID OHC 6 (165 HP)

(215-HP) SPRINT 230 CID OHC 6

108" WB

E70 x 14 W.O. TIRES

400

67 (new)

(INTRO. 2-23-67)

Firebird HO

V8 326 CID (285 HP)

HO (note STRIPE, "HO" LETTERING on SIDES. (HO means "HIGH OUTPUT")

TOTAL 1967 PRODUCTION: 82,558

CVTS. PRICED FROM $2903.

400 CID V8 (325 HP)

326 326 CID V8 (250 HP)

Firebird 326

TOTAL 1968 PRODUCTION: 107,110

(E70/F70 x 14" TIRES)

"LANDAU TOP" $84. EXTRA

AIR CONDITIONING $370.

CONVERTIBLES, $2996. UP

$3705. (WEST) (INTRO. 9-21-67)

Firebird 400.

new 250 CID FOR OHC 6 (175 HP) (SPRINT 6 = 215 HP) 350 CID V8 (265 HP) 400 CID V8 (335 HP)

CVT. (16,960 BLT.)

68

new SIDE SAFETY LIGHTS

(VARIOUS MODELS CONT'D.)

H/Ts FROM $3238; CVTS. FR. $3453 (WEST COAST PRICES)

H/T (90,152 BLT.)

H/Ts $2781. UP

400 H/T $3490. (WEST) 400 CID V8 (330 OR 335 HP)

"400" (V8) OPTION = $252. EXTRA

Firebird

H/Ts FROM $2821. (76,059 BLT.)

HOOD-MOUNTED TACH. STILL AVAILABLE (SINCE 1967)

INTRO. 9-26-68)

69 new NARROW GRILLE

$3588. (WEST)

Firebird 400 by Pontiac

CONVERTIBLES FROM $3045.

(11,649 BLT.)

Firebird NAME ON FRONT FENDER

400 CVT. $3772.

REAR SPOILER DETAILS

TRANS.-AM DASH $3196. UP

Firebird Trans Am.

69½ (new)

LONG HOOD SCOOPS

SIDE SCOOPS

WITH 400 CID V8 (335 HP) 3.55 to/ GEAR RATIO

WEST COAST: $4366. ('70 SEASON)

(3196 BLT., 1969½-70)

1969 ENGINES: 250 CID 6=(175 or 230 HP); 350 CID V8=(265, 325 or 330 HP) 400 CID V8=(330, 335 or 345 HP)

TOTAL 1969 PROD.=90,904   CALENDAR YR. PROD.=105,526

FORD

PRICES START AT **$1003.** (6 CPE.)

V8   Six

NEW OVERSIZED BRAKES (12" DRUMS)

CLUB COUPE

(70,826 BLT.)

TUDOR

FORDOR (101,302 BLT.)

SUPER DELUXE has DESIGNATION RT. of GRILLE.

TUDOR (238,324 BLT.)

There's a *Ford* in your future

1946 MODEL STARTS JULY, 1945

46

new GRILLE

NEW 1946 FORD SPORTSMAN'S CONVERTIBLE (V8) (with GENUINE WOODEN BODY)

(1209 BLT.)

$1865.

Outside and inside, there never was a car like this before!
The new Ford Sportsman's Convertible is really *two* cars in one!
Ford designers have combined the paneled smartness of the
station wagon and the touch-a-button convenience of the convertible!

114" W.B.

ALL-METAL CVT. MORE COMMONLY SEEN (ILLUSTR. ON NEXT PAGE)

PRODUCTION (OF BODY TYPES NOT ILLUSTRATED ABOVE

BUSINESS COUPE (22,919)
STATION WAGON (4 DR. with WOODEN BODY) (16,960)
CUSTOM BLT. VEHICLES ON FORD CHASSIS (123)

97 HP

1946½ **MONARCH** (INTRO. 3-23-46 IN CANADA AS A MERCURY- TYPE CAR SOLD BY CANADIAN FORD DEALERS.)

118" WB

LION FIGURE ON HUBCAP

FORD

EARLY 47
(SIMILAR TO 1946)

Ford's out Front
(1947 SLOGAN)

CONVERTIBLE (METAL)
(16,359 BLT.)

$1740.
(V8)

47½-48

1948 has NO STEERING COLUMN LOCK, AS PREVIOUSLY USED.

... new stainless steel body molding newly fashioned door handles ...

... new body colors ...

V8 WAGON

There's a FINER *Ford* in your future

A newly styled instrument panel with big new dials for easy reading

new HOOD MEDALLION IDENTIFIES 6 OR V8

MODIFIED GRILLE NO LONGER has RED INDENTATIONS.

new ROUND PARKING LIGHTS PLACED BELOW HEADLIGHTS

new wheel rims and hub caps

new heavier bumper guards—And many other new features!

1947 PRODUCTION 2-DR. (180,649); 4-DR. (116,764);
COUPE (10,872); CLUB COUPE (80,830);
WAGON (16,104) STEEL CONVERTIBLE (22,159);
SPORTSMAN CONVERTIBLE (2250);
CUSTOM CHASSIS (46)

COMBINATION IGNITION/STEERING-LOCK

1948 PRODUCTION 2-DR. (105,517); 4-DR. (71,358);
COUPE (5048); CLUB COUPE (44,826)
WAGON (8912); STEEL CONVERTIBLE (12,033)
SPORTSMAN CONVERTIBLE (ONLY 28)
MODEL YEAR ENDS MID-1948, WHEN 1949 FORDS
APPEAR.   CONVENTIONAL IGNITION LOCK, 1948

FORD

COUPE

TUDOR

(560,086 BLT.)
$1425. UP

PRICES START AT
$1333.
(DLX. COUPE)
(6 CYL.)

FORDOR
(292,739 BLT.)
$1472. UP

CONVERTIBLE
(AVAIL. ONLY IN CUSTOM)
(51,133 BLT.)
$1886.

CVT.

CUSTOM
SERIES
REPLACES
"SUPER DE LUXE."

226 CID 6 (95 HP)
and
239.4 CID V8 (100 HP)
CONTINUE.

NEW! '49
(TOTALLY RESTYLED)

STARTS
SPRING,
'48

(31,412 WAGONS BLT.) **$2119.**

Wagon

Overdrive

Engine speed 42 m.p.h.   Car speed 60 m.p.h.

(OPTIONAL)

57% more luggage space

new
"Hydra-Coil"
FRONT SPRINGS

Ford Custom

CHOICE OF COLOR
Hard Tops
1. BLACK
2. COLONY BLUE
3. BAYVIEW BLUE
4. SEA MIST GREEN
5. ARABIAN GREEN
6. MIDLAND MAROON
7. BIRCH GREY
8. GUNMETAL GREY
Convertibles
9. FEZ RED
10. MIAMI CREAM

(new CUSTOM SERIES
DESIGNATED BY NAMEPLATE
JUST AHEAD OF SIDE DOOR.)

"6" OR "8"
IN GRILLE
"SPINNER"
INDICATES
NUMBER OF
CYLINDERS.

TOTAL 1949 PROD.:

New "Flight Panel" dash ...

1,118,308
(INCLUDING
EARLY '49s
BUILT IN
1948)

FORD

"Country Squire"
STATION WAGON
(8-PASS.)

"Double Duty"

CVT.

(50,299 SOLD)

$1948.

'22,292 WAGONS BLT. 1950)

$2028.

(MOST POPULAR MODEL WAS 2-DR. SEDAN, NEARLY 700,000 SOLD!)

1,000,000 FORDS BUILT 1950; MOST FORD SALES SINCE 1930!

TOTAL SALES (INCLUDING TRUCKS) 1,208,912

50

new "KEYSTONE" FORD EMBLEM INTRODUCED

$1511.↘

CLUB COUPE

CHASSIS (V8)

(85,111 SOLD)
(ALSO 35,120 DLX. BUSINESS COUPES)

new MID-SEASON 2-DR. "CRESTLINER"

$1711.↗
(17,601 BLT.)

new EMBLEM ON HOOD (ALSO ON DECK LID)

FORD

(325,069 SOLD)

FORDOR

"TEST DRIVE" A '50 FORD

THERE'S A Ford IN YOUR FUTURE WITH A FUTURE BUILT IN!

FORD

BIG 1951 NEWS IS FORD'S FIRST H/T ("HARDTOP CONVERTIBLE,") THE *VICTORIA*!
(110,286 BLT.!) $1925.

CONVERTIBLE, BUSINESS CPE.,
CLUB CPE. and TUDOR SEDAN
ALSO AVAIL.

TOTAL 1951 PRODUCTION: 1,013,381
FORDOR (232,691 CUSTOM, 54,265 DELUXE) $1465.
(CUSTOM HAS MORE CHROME TRIM)

$1553.

$1595. ← CRESTLINER

(8703 BLT.)

51 new GRILLE

MORE PRONOUNCED TAIL LIGHT—
HOUSING EXTENSION
AROUND SIDE
OF FENDER
w. CHRM.
TRIM.

DUAL
SPINNERS
IN GRILLE

new VICTORIA H/T

VICTORIA INTERIOR

VICT. WRAPAROUND
REAR WINDOW

$2029.

COUNTRY SQUIRE
(29,017 BLT.)

new DASH

You can pay more but you can't buy better!

VICTORIA H/T with WINDOWS OPEN

1951

REAR (SEDAN)

FORD

new **MAINLINE** (has LEAST AMOUNT OF CHROME TRIM)

$1832.

(5246 BLT.)

COUNTRY SQUIRE NOW A 4-DOOR METAL WAGON has IMITATION MAHOGANY PANEL DECALS, FRAMED with REAL MAPLE OR BIRCH TRIM. $2186.

$2060.

new **RANCH WAGON** 2 DR.

1952

(32,566 BLT.)

new **COUNTRY SEDAN** 4 DR. WAGONS

(11,927 BLT.)

Station Wagons

FORD'S FIRST 4-DR. WAGONS SINCE 1948

FORD'S FIRST WAGONS W/O WOOD PANELING.

52 (TOTALLY RESTYLED)

1952 PROD.: 671,733

MAINLINE PRICED FROM $1389. $1526., WEST COAST (BUSINESS CPE.)

New Flight-Style Control Panel

Ford's new Center-Fill Fueling cuts down spillage.

$1570.

new **SUSPENDED PEDALS**

DASH

$2027.

"TEST DRIVE" A FORD TODAY YOU CAN PAY MORE BUT YOU CAN'T BUY BETTER

CUSTOMLINE 4 DR. (183,303 BLT.) $1615.

HUGE, curved, one-piece windshield and car-wide rear window to match. You can really see what's ahead and what's behind!

(22,534 BLT.)

CRESTLINE SUNLINER V8 **CRESTLINE VICTORIA** H/T

new **OHV 6** 215.3 CID 101 h.p. High-Compression

Mileage Maker Six

FORD

new **GRILLE** has APPEARANCE OF 3 "SPINNERS" $1925., OR $2104. WEST COAST

(77,320 BLT.) Full-Circle Visibility

CUSTOMLINE 2-DR. SEDAN (175,762 BLT.) SHOWN ABOVE CONVERTIBLE

"Only V8 in its field"! 110 h.p. High-Compression L-HEAD 239.4 CID V8 CONT'D. W. MORE H.P. Strato-Star V-8

FORD

CVT. IS PACE CAR AT 1953
INDY 500 RACE

PRICES START AT $1400.

($1537.
MAINLINE 6 CPE.,
(WEST COAST)

MAINLINE 2-DR
(152,995 BLT.) $1497.

DASH

CUSTOMLINE 4-DR. $1628
(374,487 BLT.)

FORD 50TH
ANNIVERSARY

VICTORIA
(128,302 BLT.)

$1941.

1953

REAR DECK DETAIL

50TH ANNIVERSARY OF FORD MOTOR CO.

3 SERIES, AS IN '52 { MAINLINE / CUSTOMLINE / CRESTLINE

53

new GRILLE

FRONT CLOSE-UP

1953 PRODUCTION:
1,247,542

ONLY ONE
"SPINNER"
IN new GRILLE.

$2203.

(11,001 BLT.)

COUNTRY
SQUIRE

SUNLINER CVT. $2043.

(40,861 BLT.)

FORD-O-MATIC
(SINCE '51)

2-DOOR
RANCH WAGON $2019. (6)
(66,976 BLT.) $1917.   (WEST COAST)

4-DOOR $2076
COUNTRY SEDAN
(37,743)

new! **Ford Skyliner** (CRESTLINE SERIES)

COUNTRY SQUIRE

NO SHIFTING...NO CLUTCHING

$2199.
(V8 $134 EXTRA)
with PLEXIGLASS
ROOF WINDOW

$2339.

New 130-h.p. Y-BLOCK V-8

239 CID V8 ENDS '54

54
new GRILLE

MAINLINE $1701.

New Ball-Joint Front Suspension

New 115-h.p. II-BLOCK SIX

223 CID (THR. '64)

UP FRONT AND BACK DOWN

4. Four-Way Power Seat.

CUSTOMLINE

4 DR. $1793.

5 optional power assists*

★ Master-Guide power steering does up to 75% of steering work . . . ★ Swift Sure Power Brakes do up to one-third of your stopping work . . . ★ Fordomatic Drive does *all* your shifting

. . . ★ Power-Lift Windows open and close at a button's touch. And ★ 4-Way Power Seat adjusts up or down, forward or back, at a touch of the controls.   *At extra cost.

PRODUCTION: MAINLINE = RANCH WAGON (44,315); 2 DR. (123,329); BUSINESS COUPE (10,665); 4 DR. (55,371)
CUSTOMLINE = RANCH WAGON (36,086); 2 DR. (293,375); CLUB COUPE (33,951); 4 DR. (262,499); 4 DR. WAGON (COUNTRY SEDAN) (48,384)
CRESTLINE = VICTORIA H/T (95,464); SKYLINER H/T (13,344); 4 DR. (99,677); SUNLINER CVT. (36,685); COUNTRY SQUIRE 4 DR., 8-PASS. WAGON (12,797)

193

FORD

PRICES START AT
$1606. (MAINLINE 6 CPE.)

$1753.

SEDAN
MAINLINE

$2224.

new FAIRLANE SUNLINER CVT. (ABOVE)

$1801. CUSTOMLINE

new GRILLE

new "WRAP-AROUND" WINDSHIELD

120-H.P. 6 OR V8s with 162 or 182 H.P.

$2043.

55

↑ RANCH WAGON

CUSTOM RANCH WAGON

$2156.
6-PASS.

COUNTRY SEDANS

"Y" SYMBOLIZES Y-BLOCK V8 new 272 CID

$2109.
$2392.
↙ COUNTRY SQUIRE

8-PASS. (with FAIRLANE SIDE TRIM)

$2287.

FAIRLANE VICTORIA

$2095.

SIDE EMBLEM (ON FAIRLANE TYPES)

FAIRLANE CROWN VICTORIA (note BAND WRAPPED OVER ROOF)
$2202.

new FAIRLANE MODELS IDENTIFIED BY SWEEP SIDE TRIM

FAIRLANE SERIES REPLACES CRESTLINE

PRODUCTION: MAINLINE = 2 DR. (76,698); BUSINESS CPE. (8809); 4 DR. (41,794)
CUSTOMLINE = 2 DR. (236,575); 4 DR. (235,417)
FAIRLANE = VICTORIA H/T (113,372); CROWN VICTORIA H/T (33,165); CROWN VICTORIA GLASSTOP
("SKYLINER" STYLE, LG. STATIONARY SUNROOF OVER FRONT SEAT AREA) (1999); 2 DR. CLUB SEDAN
(173,311); TOWN SEDAN (254,437); SUNLINER CONVERTIBLE (49,966) WAGONS = RANCH WAGON (40,493); CUSTOM
RNCH. WGN. (43,671); COUNTRY SEDAN

FORD

V-8 h.p. upped

MAINLINE — $1748. UP

CUSTOM RANCH WAG. $2249.

FAIRLANE FORDOR

1956

CTY. SQUIRE

new 2-DR. LUXURY PARKLANE WAGON (INTRO. TO COMPETE with CHEVY's NOMAD.) (RARE! 1956 ONLY) 15,186 BLT.) $2428.

MAINLINE $1748. – 1895.
CUSTOMLINE $1939. – 1985.
FAIRLANE $2047. – 2407.
WAGONS $2185. – 2533.

56 new GRILLE

1956 PRODUCTION: 1,408,478

The 272-cubic inch Ford V-8, the standard eight for all Customline and Mainline Fords. Has modern dual carburetor, automatic choke, single exhaust.

2.2.3 CID 6 INCREASED TO 137 HP @ 4200 RPM

The 292-cubic inch Thunderbird V-8, the standard eight for all Fairlanes and Station Wagons, is now available in all Customline and Mainline models, too. Has 4-barrel carburetor, dual exhausts.

202 H.P.

CUSTOM COUNTRY SEDAN

$2428. (85,374 BLT.)

$1985. 23,130 BLT.) CUSTOMLINE VICTORIA

SKYLINER   CROWN VICTORIA

The 312-cubic inch Thunderbird Special V-8,

225 h.p.

$2249. new 4-DR. H/T (FAIRLANE FORDOR VICTORIA) (32,111 BLT.)

1956 INTERIOR

FORD

6 CYL. INCREASED TO 144 HP

COUNTRY SEDAN

SQUIRE

(60,486 BLT.)

LADDER-TYPE CONTOURED FRAME

4-way ball-joint front suspension

RANCH WAGON

CUSTOM (REPLACES MAINLINE)

(116,963 BLT.)

CUSTOM TUDOR

$2451. UP

CUSTOM 300 FORDOR
(194,877 BLT.)
$2157.

$1991.

New deep-offset hypoid axle

FAIRLANE 500 MODELS BELOW

UP TO 245 HP with "SILVER ANNIVERSARY" V8s."

FAIRLANE (note UNIQUE SIDE TRIM)

TOWN SEDAN (193,162 BLT.)
(52,080 BLT.)   (FAIRLANE 500)

RESTYLED **57** (1,522,408 TOTAL PRODUCTION)

AVAIL. ENGINES

223 CID 6 (144 HP)
272 CID V8 (190 HP)
292 CID V8 (212 HP)
312 CID V8 (245 HP)

FAIRLANE 500 4-DR.
TOWN VICTORIA H/T

$2404.

LOW-SILHOUETTE CARB.

(68,550 BLT.)

(20766 BLT.)

new V8 SKYLINER has RETRACTABLE HARD TOP (POWER-OPERATED)

SUNLINER CVT.

$2942.

$2505.

(77,726 BLT.)

REAR

new FRONT END

(183,202 BLT.)

FORD

6 CYL. NOW 145 HP (THROUGH '60)

CUSTOM 300

FAIRLANE 500 SKYLINER (SHOWN *with* TOP IN MOTION, *and with* TOP IN PLACE.) **$3163.**

$2397.

RANCH WAGON (4-DR. ALSO AVAIL.)

DEL RIO RANCH WAGON **$2503.**

NOTHING NEWER IN THE WORLD

DASH

COUNTRY SEDAN **$2557.** UP

COUNTRY SQUIRE (2 VIEWS)

$2794.

NEW INTERCEPTOR V-8
PRECISION FUEL INDUCTION

A TRUE AIR RIDE

FINE-CAR DETAIL

new ROOF GROOVES

58 (TOTALLY RESTYLED)

Fairlane

4 HEADLIGHTS

$2428.

4 TAIL LIGHTS

F-1958

Versatile Cruise-O-Matic Drive! Set selector in D_1 position for brisk, solid-feeling take-off. Select D_2 for gentle, sure-footed starts. What's more, when new Cruise-O-Matic Drive is teamed with a new Interceptor V-8 engine it can give you up to 15 per cent more gasoline mileage.

V8s = 292, *new* 332 *and new* 352 CID (THROUGH '59)

Up to 300 h.p. with new Precision Fuel Induction.

Fairlane

FAIRLANE 500

PRODUCTION: CUSTOM: 2 DR. (36,272); BUSINESS SED. (4062); 4 DR. (27,811); CUSTOM 300: 2 DR. (137,169); 4 DR. (135,557); FAIRLANE: VICT. H/T SEDAN (5868); 4 DR. TOWN SEDAN (57,490); VICT. H/T (16,416); 2 DR. CLUB SED. (38,366) FAIRLANE 500: SKYLINER RETRACTABLE H/T (14,713); VICT. H/T (36,509); 4 DR. (105,698); VICT. H/T (80,439); " " (34,041); SUNLINER CVT. (35,029) WAGONS: 2 DR. RANCH WAGON (34,578); 2 DR. DEL RIO (12,687); 4 DR. RANCH WAG. (32,854); COUNTRY SEDAN (68,772); (COUNTRY SEDAN, 9-PASS. (20,702); COUNTRY SQUIRE, 9-PASS. (15,020) (1,038,560 1958 TOTAL)

FORD

CUSTOM 300

$2132. (6)

WINDSHIELD DOGLEG DETAILS

note "FORD" LETTERING ON HOOD

(23,892 BLT.)

Fairlane 500

$2537.

FAIRLANE 500 VICTORIA ROOFLINE (CLOSE-UP)

note EMBLEM ON HOOD

new GRILLE

FAIRLANE REAR FENDER

(TOTALLY RESTYLED AGAIN)

TOTAL 1959 PRODUCTION: 1,427,835

59

ENGINES:
223 CID 6 (145 HP)
292 CID V8 (200 HP)
332 CID V8 (225 OR 300 HP)

1959 WAGONS ON NEXT PAGE

GALAXIE new MID-SEASON SERIES

NEW FORD GALAXIE CLUB VICTORIA—THUNDERBIRD STYLING IN A 6-PASSENGER, 2-DOOR HARDTOP

(121,869 BLT.)

$3346.

(12,915 BLT.)

new 1959½ TOP-OF-LINE GALAXIE MODELS ADDED, with T-BIRD ROOFLINE.

THE FINAL SKYLINER (GALAXIE)

FORD

$2634.

wagons

4-DOOR
and
2-DOOR
RANCH
WAGONS

FENDER
CHEVRONS
ON THIS '59½
RANCH WAGON

(INTERIOR VIEW EXAGGERATED)

ROOMY NEW FORD RANCH WAGON . . . LOWEST PRICED WAGON OF THE MOST POPULAR THREE

STATION WAGON PRODUCTION

59 (CONT'D.)

59-C 2-DR. RANCH WAGON (45,588)
59-D " " DEL RIO (8663)
71-E COUNTRY SEDAN, 9-PASS. (28,811)
71-F " " , 6-PASS. (94,601)
71-C COUNTRY SQUIRE, 9-PASS. (24,336)
71-H 4-DR. RANCH WAGON (67,339)

$2958.   (WEST COAST)   **$3076.** *

COUNTRY
SQUIRE

1959 DASH
(ILLUSTR. with
FACTORY-INSTALLED
AIR CONDITIONER
UNIT)

*WAGON PRICE
SHOWN APPLIES TO
V-8 9-PASSENGER
6 CYL. or
6-PASS. MODELS
also avail.

COUNTRY
SEDAN
$2947. *
(WEST COAST)
(ELSEWHERE, $2745. OR $2829.)

FORD
(91,041 BLT.)

FAIRLANE 500

FAIRLANE PRICES START AT **$2170.**
(6-CYL. 2-DR.)

(31,866 BLT.)

GALAXIE TUDOR

(68,461 BLT.)
new STARLINER
$2723. (V8; 6 ALSO AVAIL.)

ARCHED TAIL-LIGHTS ONLY ON 1960 MODELS →

(TOP UP)

SUNLINER CVT.

(TOP DOWN)

292 cid V8 REDUCED TO 185 HP; OTHER ENGINES AS IN 1959.

$2860.

(44,762 BLT.)

GALAXIE FORDOR
$2603.

(104,784 BLT.)

(TOTAL 1960 PRODUCTION = 1,004,305)

NEW SLOPING HOOD GIVES INCREASED VISIBILITY

60 (TOTALLY RESTYLED FOR 3RD SUCCESSIVE YEAR!)

145 TO 300 HP

DASH

(27,136 BLT.)

RANCH WAGON

$2586.

COUNTRY SEDAN →
(59,302 BLT. 6 PASS.)
(19,277 BLT. 9 PASS.)

$2967.

9-passenger Country Squire (22,237 BLT.)

Beautifully built to take care of itself...

FORD WHEEL COVER

(66,875 BLT.)

2 DR.
FAIRLANE
$2261.
(6)

$2432.

4 DR.
(98,917 BLT.)
FAIRLANE 500

new GRILLE IS CONCAVE, BISECTED HORIZONTALLY

61

RETURN TO CONSERVATIVE STYLING

$2664.
(30,342 BLT.)
GALAXIE 4-DR. TOWN VICTORIA H/T

(CLOSER VIEW of GALAXIE WHEEL COVER at UPPER RIGHT)

$2588.
(12,042 BLT.)
RANCH WAGON

$2754. UP

STATION WAGONS.

CNTRY. SEDAN

292,352 OR new 390 CID V8s (175 TO 401 HP)

SQUIRE

(16,961 BLT.)
$2943.

GALAXIE VICTORIA H/T

(CLOSE-UP and DASH)

$2599.

(29,669 BLT.)
STARLINER H/T
$2599.

ROUND TAIL LIGHTS RETURN

1961 PRODUCTION = 1,338,790

(75,437 BLT.)

201

FORD

(NO MORE 2-DR. WAGONS)

RANCH WAGON (33,674 BLT.) $2733.

(47,635 BLT.)

COUNTRY SQUIRE (16,114 BLT.)

6-PASSENGER COUNTRY SEDAN (9-pass. model also) $2829.

$3018.

"CRUISE-O-MATIC" A/T CONTROL

SLOGAN: Live it up with a lively One from FORD

("500") (27,824 BLT.)

(54,930 BLT.) $2453.

Galaxie

new BLUNTED REAR END

XL H/T (28,412 BLT.)

138 to 405 HP (THR. '63)

62

POWER STEERING

GALAXIE 500/XL.

DENOTES 405 HP THUNDERBIRD ENGINE

GALAXIE 500 and XL have GRILLE MEDALLION

1962

(87,562 BLT.) *Galaxie* 500

1962 TAIL LIGHTS

BUCKET SEATS and FLOOR CONSOLE IN new *Galaxie* 500/XL!

new ✳ SIDE TRIM

$3518.

Galaxie 500 XL

(13,183 BLT.)

1962 PRODUCTION: 1,476,031

$2924. (42,646 BLT.) *Galaxie* 500 (SEDAN and CVT. ILLUSTR.)

✳=ON 500, XL

202

FORD

GALAXIE 500

$2674.

$2739.

(39,154 BLT.)

SQUIRE

$2924.

(29,713 BLT.)

63

new GRILLE with SHIELD EMBLEM, AND STEP-UP ALONG LOWER EDGE

new SIDE TRIM

note INDENTATION ALONG UPPER BORDER OF WOOD-LIKE "COUNTRY SQUIRE" SIDE TRIM.

1963 PRODUCTION: 911,496 (FULL-SIZE)

SQUIRE (6 PASS.) (20,359 BLT.) $3018.
(9 PASS.) (19,567 BLT.) $3088.

COUNTRY SEDAN WAGONS ALSO =
(6 PASS.) (64,954 BLT.) $2829.
(9 PASS.) (22,250 BLT.) $2933.

ENGINES	223 CID 6 (138 HP)
	289 CID V8 (195 OR 271 HP)
	352 CID V8 (250 HP)
	390 CID V8 (300 OR 330 HP)
new 427 CID V8 (425 HP)	

BACKGROUND: MONACO, ON THE RIVIERA

FLOOR CONSOLE and DASH

PRICES START AT $2563.
6-CYL. "300" 2-DR.

UP TO 425 HP IN *new* '63½

new

Presenting the 63½ Super Torque Ford Sports Hardtop —brand new hardtop that looks like a convertible!

new SWING-AWAY STEERING WHEEL AVAILABLE

CLEAR GLASS BACKLIGHT IN CVT.

RANCH WAGON RETURNS

CUSTOM RANCH WAGON

FORD

SQUIRE

138 TO 425 HP

GALAXIE 500 4-DR. H/T (49,242 BLT.)

$2750.

$3988. UP

1964

GALAXIE 500 XL

1964

new GRILLE

64

1964

PRICES START AT $2586.

($2600 IN '65)

(CUSTOM 6 2-DR.)

TOTAL 1964 PRODUCTION = 881,061 (FULL-SIZE)

CUSTOM 4 DR. (96,393 BLT.)

$2518.

TOTAL 1965 PRODUCTION = 1,048,388 (FULL-SIZE)

"CRUISE-O-MATIC" A/T CONTROL

$2361. ~ 3498. PRICE RANGE

(RESTYLED)   150 TO 425 HP (THROUGH '67)

65

CUSTOM 500

(71,727 BLT.)

$2415.

P R N DRIVE L

new VERTICAL STACKED HEADLIGHTS →

new DIP IN WAGON ROOF

1965

Convenient face-to-face rear seats add passenger space

CVT. (9849) BLT.

$3104. UP

SQUIRE

GALAXIE 500 XL

$3498.

$3313.

GALAXIE 500 LTD

(68,038 BLT.)

FORD

1965

TAIL LIGHT SHAPE IS new

FORD $2533.

RANCH WAGON
(33,306 BLT.)
$2793.

CUSTOM 500

COUNTRY SEDAN

GALAXIE 500
240 cid 6 (150 HP)
289, 352, 390,
427 and
428 cid
VBs AVAIL.
(200 to
425 HP)

swings down for cargo
$2882. UP

$2685. UP
(7 LITRE ENG.,
$3621.)

66

new
2-TIERED
GRILLE

119" WB (THROUGH '68)

COUNTRY
SQUIRE

'69,598
BLT.)

Swings open for people

$3182. UP

$3243. 4-DR. H/T

XL

CUSTOM
500

240 cid 6
(150 HP)
289 cid V8
(200 HP)
390 cid V8
(315 HP) ALSO,
427/428 VBs
(TO 345 HP)

GALAXIE 500
H/T ROOFLINE

LTD

67

H/T
$3362.

83,260
BLT.)
$2595.

1967

$2441.~3493. PRICE RANGE

COUNTRY
SQUIRE

$3234.
UP

New Magic Doorgate for all our 1966 wagons!

205

FORD
$3048.

CUSTOM (CUST. 500, GAL. 500 ALSO USE THIS GRILLE)

XL INTERIOR

(INTRO. 9-22-67)

68

new GRILLES
302 CID V8 (new)
(210 HP)
6 ALSO CONT'D.

DUAL-FACING REAR SEATS AVAIL.

COUNTRY SQUIRE
(LTD)
$3977.

XL H/T

428 CID V8
(345 HP)
OPT. ON ALL MODELS

new CONCEALED HEADLIGHTS (LTD, XL)

(SHOWN OPEN)

$3257.

new 112" WB (THROUGH '78)

DASH

new 150 HP IN 6

CUSTOM 500
SIDE CHROME STRIP NOT SEEN ON CUSTOM.)

(INTRO. 9-27-68)

69

new GRILLES

$3514.

LTD

LTD INSIGNIA ABOVE GRILLE

XL
SPRTSRF. H/T
$3536

It's the going thing!

XL

206

FRAZER

KAISER-FRAZER CORPORATION
• WILLOW RUN, MICHIGAN

(1946 – 1951)

123½" W.B.
THROUGH '51)

(REPLACES PRE-WAR GRAHAM.)

47

F-47

EARLY FRAZERS
(BLT. 1946)
have PAINTED GRILLE.

LATER MODEL,
with CHROME GRILLE →

EMBLEM

JE SUIS PRET

100 HP @ 3600 RPM   7.3 COMPRESS.

(112 HP OPTIONAL "MANHATTAN")

6 CYL., L-HEAD CONTINENTAL ENGINES   226.2 CID
(USED IN ALL FRAZERS)

F47 = STANDARD SEDAN (36,120 BLT.) - $2295.
F47C = MANHATTAN   "   (32,655 BLT.) - $2712.

F485(1) = STANDARD SEDAN (29,480 BLT.) - $2483.
F486(1) = MANHATTAN   "   (18,591 BLT.) - $2746.

(SAME ENGS. AS 1947)

SEE ALSO:
KAISER

3⁵⁄₁₆" × 4³⁄₈" BORE and STROKE
(KAISER SPECS. SIMILAR)

FRAZER REAR VIEW

new = 4 FRONT BUMPER GUARDS

(EARLIEST '48s SOMETIMES CONSIDERED "1947½.")

48

F-485 ; (MANHATTAN SEDAN IS NOW F-486)

ILLUSTRATED
ON FAMOUS "17-MILE-DRIVE,"
AT PEBBLE BEACH, CALIF.

ALL
FRAZERS
ARE
4-DOOR
MODELS.

FRAZER

JOSEPH W. FRAZER (left) and HENRY J. KAISER (right) STANDING BY THE 200,000th CAR (A 1948 FRAZER) TO BE BUILT BY THE KAISER-FRAZER CORP.

200,000th

J.W. FRAZER

H.J. KAISER

112 HP (ALL)

$2595.

VAGABOND (HATCHBACK!) PRICED FROM $2321.

F-505 and F-506 MANHATTAN are 1950 MODELS.

MANHATTAN SEDAN has HEAVY BAND of SIDE CHROME

(9950 ? BLT.)

(24,923 TOTAL '49~'50 PROD.)

49-50

1949 = F-495 or F-496 MANHTN

new LARGE GRIL and PARKIN LIGHTS

(1950 MODE ENDS 2-50)

H.P. INCREASED TO 115

(RESTYLED) F-515 OR F-516 MANHATTAN

51

(1951 MODELS START FEB., 1950)

F 515 STANDARD
5151 SEDAN (6931 BLT.?) ;   5155 VAGABOND (3000 BLT.

F 516 MANHATTAN
5161 4-DR. H/T (152 BLT.)
5162 4-DR. CONVERTIBLE (131 BLT.)

4-DOOR CVT. SEDAN

new "MANHATTAN" LOOKS LIKE THE 4-DR. CVT., BUT HAS STEEL PAINTED OR NYLON-PADDED TOP SECTION. (SAME PRICE AS CVT. SEDAN)

VAGABOND

$3075.

DASH (SIMILAR TO 1949)

$2359.

$2399.

STD. SEDAN

EXPERIMENTAL SAFETY CAR BLT. 1945 TO 1947 BY
H. GORDON HANSEN,
AT SAN LORENZO, CALIF.

FORD V8 ENGINE

GORDON DIAMOND

156" WB BETWEEN FRONT REAR
SINGLE WHEELS. ANOTHER PAIR OF
WHEELS "AMIDSHIPS."
PURCHASED BY HARRAH'S AUTOMOBILE COLLECTION

GRAHAM (and HUPMOBILE)

6 - CYL.
L - HEAD
ENGINES

115"
WB

40-41

'41 GRAHAM:
$895. and up

'40 HUPP:
$1145. and up

FORMER CORD
BODY DIES
USED

GRAHAM "HOLLYWOOD" and
HUPMOBILE "SKYLARK" LOOK ALMOST ALIKE!

THOUGH THESE ARE PRE~1946 CARS, THEY ARE
INCLUDED BECAUSE OF GRAHAM (GRAHAM~PAIGE CORP.)
ANCESTRY TO POSTWAR KAISER-FRAZER CORP.

GREGORY

(1949)   (ANOTHER EXPER. MODEL, 1952)

BEN GREGORY, MFR.,
KANSAS CITY, MO.

4-CYL. Continental
REAR ENGINE
FRONT-WHEEL-
DRIVE

49

PRODUCTION ATTEMPTED

40 HP
94" WB

$1050. (PROPOSED PRICE)

(1948-1949)

HOPPENSTAND

HOPPENSTAND MOTORS, INC., GREENVILLE, PA.

2-CYL. FLAT, AIR-COOLED, REAR ENG.

48-49

(FULL PRODUCTION NEVER ACHIEVED.)

HENRY J (KAISER-FRAZER CORP.) FOLLOWS →

(1950-1954)

Henry J

PRICED FROM
$1299.
(WITH PERIODIC INCREASES)

4 OR 6-CYL.
"SUPERSONIC"
ENGINES

(2-DRS.
ONLY)

THIS
GRILLE
STYLE
RETAINED
ON 1952
HENRY J

K 513 = 4 CYL.
K 514 = DELUXE 6 CYL.

51
(INTRO. 1950)

STD.
(38,500 BLT.)
DLX.
(43,400 BLT.)

KAISER-FRAZER CORPORATION, WILLOW RUN, MICHIGAN

GIVEN
THE
FASHION
ACADEMY
GOLD MEDAL
AWARD

ALLSTATE PRODUCTION

1952 (4 CYL.) 110 BASIC (200)
111 STD. (500); 113 DELUXE (200)
(6-CYL.) BASIC (200); " (466)
1953 (4 CYL.) STD. (200);
DELUXE (225)
(6 CYL.) DELUXE (372)

('52)

ALLSTATE CAR = SPECIAL SERIES
OF HENRY J, SOLD EXCLUSIVELY
BY SEARS, ROEBUCK and CO.

134.2 CID 4 (68 HP) OR
161 CID 6 (80 HP)
(SAME ENGINES ALLSTATE OR
1951 TO 1954 HENRY J)

52-54

HENRY J
CORSAIR
PRODUCTION (APPR.)
1952 (7600); DLX. (8900)
1953 (8500); DLX. (8100)
1954 (800); DLX. (300)

$1395. TO
$1785.

$1517.
(CORSAIR, '52)

CORSAIR
('53-
'54)

$1407. UP
VAGABOND 1952 ONLY

new
VAGABOND
has REAR
"CONTINENTAL" SPARE
TIRE/WHEEL

VAGABOND (3000 BLT.)
VAG. DLX. (4000 BLT.)

(*'52*) *Henry J* Vagabond

new
GRILLE
ON
'53-54
1953 TOTAL
(16,672)
1954 TOTAL
(1123)

UDSON

POSTWAR PRODUCTION RESUMES
8-30-45. 5,005 BLT. 1945;
93,870 BLT. 1946

121" WB ON ALL (THROUGH '47)
PRICED FROM
$1481.

SUPER 6

BROUGHAM CVT.
(1035 BLT.)
$1879.

46

COMM. has
2 VERT.
STRIPS
ON REAR
WINDOW

new GRILLE with RECESSED
CENTER SECTION

COMMODORE (has
HUDSON TRIANGLE EMBLEM AT FRONT END OF CHROME BELT STRIP)

ENGINES (SINCE 1946) =
2/2 CID 6 (102 HP)
OR 254 CID STRAIGHT-8 (128 HP)

SUPER 6 = $1628. up   SUPER 8 = $1855. up
COMMODORE 6 = $1887. up
COMMODORE 8 = $1955. up

$1896.

SUPER 6
SEDAN

$1749.

47

103,310
BLT. 1947

SIMILAR TO 1946, BUT has
HEAVIER CHROME MOULDING MARGIN
AROUND MEDALLION OVER GRILLE.

COMMODORE 6
SEDAN

MODELS = SUPER 6 (49,388 BLT. '48; 91,333 BLT. '49) 4 DR. SED.; 2 DR. BROUGH.; CPE.; CLUB CPE.; CVT.
COMMODORE 6 (27,159 BLT. '48; 32,715 BLT. '49) 4 DR. SEDAN; CLUB COUPE; CONVERTIBLE
SUPER 8 (5338 BLT. '48; 6365 BLT. '49) 4 DR. SEDAN; CLUB COUPE; ('49 = 2-DR. BROUGHAM)
COMMODORE 8 (35,315 BLT. '48; 28,687 BLT. '49) 4-DR. SEDAN; CLUB COUPE; CONVERTIBLE

4,119 BLT. 1948

2,462 BLT. 1949

CVTS. NOW HAVE MORE
METAL ABOVE WINDSHIELD

new "Step Down" BODIES
SURROUNDED BY FRAME

INTERIOR
('49)

48-49

(TOTALLY
RESTYLED)

(BIGGEST CHANGE IN HUDSON HISTORY)

new
124" WB ON ALL

"This time it's *Hudson*"

WITH "The New Step-Down Ride"

Hudson

new PACEMAKER 6 is
LOWER-PRICED SERIES
(119" WB)

$1933.

REAR SEAT VIEW

ROAD CLEARANCE

50

INVERTED
"V" ON
new
GRILLE

• NOW—3 GREAT SERIES • LOWER-PRICED PACEMAKER • FAMOUS SUPER • CUSTOM COMMODORE

SEDAN

COMMODORE
$2282.
$2360
(8)

Hudson is the only motor car with a recessed floor ("step-down" design). This results in the lowest-built car of them all, with true streamlining and magnificent beauty. It provides full road clearance and the most room in any automobile at any price! It creates America's lowest center of gravity, which brings you the best and safest ride ever.

124" WB
ON ALL
MODELS
EXCEPT
PACEMAKER and
PACEMAKER
DELUXE

ENGINES:
232 CID 6 (112 HP) (PACEMAKER)
262 CID 6 (123 HP)
254 CID STRAIGHT-8 (128 HP)

143,586 BLT. 1950

(COUPE = $1965.)

PACEMAKER 6

$2642. CVT. (430 BLT.)

92,859 BLT. 1951

51

$2238.

SUPER 6

2-DR. BROUGHAM

COMMODORE 8
CLUB COUPE
$2543.

new
GRILLE
(HEAVIER and
ARCHED)

$2568.

new
HORNET 6 = WITH new
308 CID 6
(145 HP)
"HORNET" ENGINE
(3 OTHER HUDSON
ENGINES
CONT'D. FROM '50)

"Monobilt"

BODY/FRAME CONSTRUCTION

HUDSON
YOUR BEST BUY
FOR THE LONG TOMORROW

212

$2264.

PACEMAKER 6   B.ROUGHAMS

COMMODORE 6

CVT. (120 BLT.)

3247.

742.
HORNET CLUB CPE.

SEDAN

79,117
BLT. 1952

new HUDSON WASP with 6-CYL. "H-127" ENG.

Hollywood H/T (new)

$2812.

HUDSON WASP TWO-DOOR BROUGHAM $2413.

HOLLYWOOD WASP   (1320 BLT.)

CLUB CPE. $2466.

new lower-priced running mate

52

HUDSON HORNET

SEDAN
$2789.

HORNET

HUDSON

B-W
ENGINEERING PRODUCTION

equipped with
B-W OVERDRIVE!
(OPTIONAL)

HYDRA-MATIC DRIVE
available for all '52 Hudsons
at extra cost.

ENGINES :
232 CID 6 (112 HP) (PACEMAKER)
262 CID 6 (127 HP) (WASP, CMDR. 6)
308 CID 6 (145 HP) (HORNET)
254 CID STRAIGHT 8
(128 HP) (COMMODORE 8)

Hudson-Aire Hardtop Styling
at standard sedan and coupe prices

COMM. 8 CONVERTIBLE
(ONLY 30 BLT.)

$3342.

COMMODORE 8   FINAL YEAR FOR STRAIGHT 8

HUDSON

2-DR. SEDAN

53

ENGINES :
232 CID 6 (127 HP) (WASP)
308 CID 8 (145, 160 OR 170 HP)
(HORNET)

HORNET
SEDAN
$2769.

SUPER WASP $2413.
119" WB

1953 PRODUCTION :
4C WASP ; 5C SUPER WASP (17,792)
7C HORNET (27,208)

6-CYL.
MODELS ONLY
(THROUGH
1954)

new
HOOD
"AIR-SCOOP"
and new GRILLE
w/o INVERTED "V."

124" WB

(The JETS
SHOWN ON "JET" PAGE)

The **WASPS** $2466.
in the low-medium
price field

SUPER WASP

new HIGH
TAIL-LIGHTS

**HUDSON DIVISION OF
AMERICAN MOTORS**
(RESULT OF MERGER WITH NASH,
5-1-54)

new 1-PC.
WINDSHIELD

The **HORNET**
in the medium price field
$2769.

CLUB COUPES

$2619.

new FRONT END
DESIGN

54
(RESTYLED)

NEW HORNET SPECIAL
available in Four-Door Sedan, Club Sedan
and Club Coupe—all at new low prices

$2571.

new 262 CID 6 ADDED
(140 HP) SU. WASP

HORNET has
160 HP

2-DR.
CLUB
SEDAN

(170 HP with
"Twin H" Power)

4 DR.
SEDAN

INTERIOR of HOLLYWOOD
(CAR ILLUS. NEXT PAGE)

$2619.

new CHROME PC. ON SIDE →

HUDSON
54
(CONT'D.)

$2988.

HORNET HOLLYWOOD H/T

32,293 HUDSON CARS BLT. 1954

52,688 BLT. 1955 ("HUDSON" NAME also USED on SOME Ramblers and Metropolitans)

new SHORTER WHEELBASES = WASP 114.3" HORNET 121.3"

new ENGINE CHOICES

V8 CHAMPIONSHIP 6

CUSTOM WASP SEDAN $2460.

55

(TOTALLY RESTYLED with NASH BODY DESIGN)

HOLLYWOOD H/T

320 CID PACKARD V8 USED

new PEAKS OVER HEADLIGHTS

22,588 BLT. 1956

HOLLYWOOD H/T

BIG new DIAMOND-SHAPED GRILLE

56

SEDAN

DASH

ENGINES: 202 CID 6 (120 or 130 HP) (WASP).
308 CID 6 (165 or 175 HP) (HORNET)
320 CID V8 (208 HP) (HORNET V8, to MARCH, 1956)
250 CID V8 (190 HP) ("   ", MARCH, 1956 on)

WASP SEDAN (2519 BLT.) HORNET SPECIAL SED. (1528 BLT.); H/T (229 BLT.); HORNET SUPER, CUSTOM SEDANS (3022 BLT.) H/T (358 BLT.) HORNET V8 SEDAN (1962 BLT.); H/T (1053 BLT.)

$2214. - $3159. PRICE RANGE

HUDSON

DASH
with new Hydra-Matic

327 CID
World's newest V-8 . . . 255 hp

HORNET SUPER

new SIDE TRIM MOULDINGS

Hornet V-8
121.3" W.B.

Lower outside by 2 full inches

IS ONLY AVAIL. MODEL (SUPER OR CUSTOM)

57
new "V" EMBLEM ON GRILLE

ONLY 4,080 BLT. 1957*
(OTHER SOURCES SAY 3876 BLT., OR ONLY 1345)

ONLY 1 V8 ENG. AVAIL.
(NO 6-CYL. MODELS)

HORNET HOLLYWOOD H/T
(APPEARS LONGER IN ← PHOTO AT LEFT THAN IN PHOTO ABOVE)

Slim outside for easy maneuvering

(DISCONTINUED JUNE 25, 1957)

. . . way up in power, way down in price!

357-1 **HORNET SUPER** SEDAN $2821.
HOLLYWOOD H/T $2911.
357-2 **HORNET CUSTOM** SEDAN $3011.
HOLLYWOOD H/T $3101.

* HIGHER PRODUCTION FIGURES MAY INCLUDE "HUDSON" RAMBLERS and METROPOLITANS.

IMP

INTERNATIONAL MOTOR PRODUCTS CO., GLENDALE, CALIF.

49-50

(1949-50)

FIBERGLASS BODY
63" WB   APPR. 475 lbs.
1-CYL., 7-H.P. GLADDEN engine

LIST FINAL DATE AS 1955.

SOME REPORTS

IMPERIAL

CUSTOM
(C-64)
133½" WB
$4260.

$324
BLT.)

EARLIER MODELS ILLUSTRATED
(WITH CHRYSLER.)

IMPERIAL CONSIDERED A TOP-
LINE CHRYSLER SERIES,
1926 TO 1954.

331.1 CID V8 (3¹³⁄₁₆ × 3⅝)
235 HP @
4400 RPM

54

CROWN (C-66)   145½" WB

7.5 COMPR.
(SINCE '51)

| 8-PASS. SEDAN (23 BLT.) | $6922. |
| 8-PASS. LIMOUSINE (77 BLT.) | $7044. |

SEDAN (7840 BLT.)
$4483.

$4720.

NEWPORT H/T
(3418 BLT.)

new
331 CID V8 (3.81 × 3.63)
250 HP @
4600
RPM

55

IMPERIAL (C-69) 130" WB

CROWN IMPERIAL (C-70)
149½" WB (THROUGH '56)
(IMPERIAL CONSIDERED AN
INDIVIDUAL MAKE, AS OF 1955.)

new 8.5 COMPRESSION

IMPERIAL
(C-73)
new 133" WB

SEDAN
(6821 BLT.)
$4832.

354 CID; 280 HP @
4600 RPM

SOUTHAMPTON H/T
(2094 BLT.)

$5094.

56

$7603. UP
CROWN IMPERIAL
(C-70)

SOUTHAMPTON
4-DR. H/T
ALSO AVAIL.

IMPERIAL

(IM1-2) CROWN $5598. CVT. (1167 BLT.)   new 129" WB (THROUGH '66)

new 392 CID (THROUGH '58) 325 HP @ 4600 RPM new 9.25 COMPR.

57

new 129" WB (THROUGH '66)

note DIFFERENCES IN NUMBER OF HEADLIGHTS

SEDAN (1729 BLT.)

SEDAN (3642 BLT.)   CROWN (IM1-2) $5406.

LE BARON (IM1-4) $5743.

LY1-L IMPERIAL   SOUTHAMPTON H/T (1801 BLT.); SOUTHAMPTON 4-DR. H/T (3336 BLT.); SEDAN (1926 BLT.)
LY1-M CROWN   "   " (1939 BLT.);   "   " " " (4146 BLT.); " (1240 BLT.);
CONVERTIBLE (675 BLT.)   LE BARON   "   " " " (538 BLT.); " (501 BLT.)
(LY1-H)   CROWN IMPERIAL LIMO. (31 BLT.)
($15,075.)

(538 BLT.)
LE BARON SOUTHAMPTON

new 10.0 COMPRESSION 345 HP @ 4600 RPM

FENDER-GRILLE DETAILS

LY1 SERIES

58

$5969.

IMPERIAL NAME (NON- LE BARONS)

218

IMPERIAL

CUSTOM SOUTHAMPTON (MY1-L)

$6103.

4910.

(1743 BLT.)

59

LE BARON SOUTHAMPTON (622 BLT.) (MY1-H)

LE BARON

new CROWN (MY1-M).
413 CID, 10.1 COMPR. (THROUGH '65) 350 HP @ 4600 RPM (THROUGH 61) '65)

(ALSO 392 CID V8 (325 HP) IN CROWN IMPERIAL)

$4910. - $6103. PRICE RANGE

ALSO, 7 CROWN IMPERIAL LIMOS.
AT $15,075. EA.

(PY1-M) CROWN

SEDAN (1594 BLT.)
$5647.

60

(PY1-L) CUSTOM SOUTHAMPTON

$4933.
TO
$6318.
PRICE RANGE

CUSTOM 4-DOOR SOUTHAMPTON (3953 BLT.)

(1498 BLT.)
$4923.

$5029.

new
8.20 x 15 TIRES
(THROUGH '64)

(PY3-H) LE BARON

IMPERIAL

CROWN
(RYI-M)
CVT.
(429 BLT.)

129"
WB

$5774.

SOUTHAMPTON

new "FREE-STANDING, HEADLIGHTS (THROUGH '63)

61

new GRILLE

$4923.-$6426. PRICE RANGE
9 CRN. IMPL. LIMOS. BLT. $16,500.
(149½"WB)

(RYI- SERIES)

America's Most Carefully Built Car

SYI-L CUSTOM #912 SOUTHAMPTON H/T (826 BLT.); #914 SOUTHAMPTON 4-DR. H/T (3587 BLT.)
SYI-M CROWN #922 SOUTHAMPTON H/T (1010 BLT.); SOUTHAMPTON 4-DR. H/T (#924) (6911 BLT.); CVT. (#925) (534 BLT.)
SYI-H LE BARON #934 " 4-DR. H/T (1449 BLT.)

$4920. - $6422. PRICE RANGE   129" WB

$6422.

ORNAMENT at HOOD
FRONT; new
SPLIT
GRILLE

IMPERIAL LE BARON 4-DR. SOUTHAMPTON (SYI-H)

(CUSTOM
is SYI-L)

62

CROWN
(SYI-M) $5400.

RAISED
TAIL-LIGHTS

HP REDUCED TO
340 @ 4600 RPM
(THROUGH '65)

two-door Southampton

4-DR.
SOUTHAMPTON
$5644.

IMPERIAL

CROWN (TYI-M) $5656.

WHEEL COVER

4-DR. H/T (6960 BLT.)

HIGH, NARROW TAIL-LIGHTS

$5058.

(FINAL YEAR FOR "CUSTOM" SERIES.)

DASH

(TYI-L) CUSTOM

63

(TYI SERIES)

(HAND-BUFFED ACRYLIC ENAMELS)

H/T (749 BLT.)

(TYI-H)

IMPERIAL LeBARON

LeBARON Coachwork

The LeBaron cloisonné crest on the roof makes this the only car on which this federal jewelry excise tax is paid.

FREE-STANDING HEADLIGHTS FOR 3RD AND FINAL YEAR

CLOSE-UP VIEW OF FRONT END (NEW GRILLE)

$5058. – $6434. PRICE RANGE

CROWN IMPERIAL LIMOUSINE (13 BLT.)
$18,500.

IMPERIAL

CROWN COUPE
(VYI-M)

$5581.

(VYI-M)

$5739.

Imperial Crown 4-Door Hardtop

LE BARON
(VYI-H)

CROWN CVT.

$5770.

DASH

(VYI SERIES)

64

(TOTALLY RESTYLED; new DESIGN
SOMEWHAT RESEMBLES LINCOLN CONTINENTAL.)

EAGLE CREST on
LE BARON VINYL TOP →

AUTO PILOT (left)
AM/FM RADIO (above)

HEADLIGHTS MOVED into
new SPLIT GRILLE.

The Incomparable IMPERIAL

VYI-M CROWN H/T (5233 BLT.); 4-DR. H/T (14,181 BLT.); CONVERTIBLE (922 BLT.)
VYI-H LE BARON 4 DR. H/T (2949 BLT.)
CROWN IMPERIAL LIMOUSINE (10 BLT.) $5581.—$6455. PRICE RANGE
        $18,500.   ($5865. to $6740., WEST COAST)

222

IMPERIAL

CROWN COUPE (AYI-M)

4-DR. H/T

$5772.

$5930.

REAR DETAILS

CHOICES OF UPH. INCL. REAL LEATHER

65 AYI SERIES

← DASH

...IGHT FLASHES IF FUEL, OIL, ...EMP. GAUGES NEED ATTENTION.

...IR COND. DUCTS (ON DASH)

HEADLIGHTS PAIRED BEHIND GLASS PANELS.

new GRILLE

(LE BARON IS AYI-H)

AYI-M **CROWN** H/T (3974 BLT.) ; 4-DR. H/T (11,628 BLT.) ; CONVERTIBLE (633 BLT.)
AYI-H **LE BARON** 4-DR. H/T (2164 BLT.)
$5772. — $6596. PRICE RANGE

ALSO, 10 **CROWN IMPERIAL** LIMOUSINES, $18,500. EA.

(1926-1975 ; 1981-) IMPERIAL

BY CHRYSLER CORPORATION

$5887.

129" WB

(5.14 BLT.) $6164.

(2373 BLT.)

$6505. (WEST COAST)   CROWN CPE. (H/T)

V8 ENGINE (SINCE '55)
new 440 C.I.D
(350 HP)
9.15 x 15 TIRES

66 BY3

BY3-M
CROWN SERIES
and   BY3-H
$7158. LE BARON
4-DR. H/T

new GRILLE

$5733.

CONVERT. AVAIL. AT $6764. (WEST COAST)

(WEST COAST) FR. $6351. (4-DR. H/T

(8977 BLT.)

THE INCOMPARABLE **IMPERIAL**
Finest of the fine cars built by Chrysler Corporation

(2193 BLT.) $5374.

IMPERIAL NAME SET IN new ALL-HORIZONTAL GRILLE

Imperial '67... the newest prestige automobile in a decade

← new 4-DR. SEDAN ADDED $5991. (WEST COAST)

new CORNER LTS. IN FENDERS

new SHORTER 127" WB

67 CYL

(INTRO. 9-29-66)

CVT. (577 BLT.)
$6861. (WEST COAST)

$6244.

IMPERIAL CONVERTIBLE DISCONTINUED AFTER 1968

REAR

LE BARON LIMO. (163" WB) WITH STAGEWAY BODY (6 BLT.) $15,000.

TOTAL 1966 PRODUCTION = 13,742   1966 CALENDAR YEAR : 17,653
  "   1967   "   = 14,620 $5733.~6540. 1966 PRICE RANGE
                                         5374.~17,000. 1967 "   "

IMPERIAL

If you want more than luxury in your luxury car.

CROWN
COUPE
$5722.
(656 BLT.)

$6381.
(WEST COAST)

68

$6940.

(1852 BLT.)

LE BARON
$7599.
(WEST COAST)

(INTRO. 9-14-67)

(FINAL CVT.)

440 cid V8 (350 HP)

(360 HP WITH
DUAL EXHAUSTS)

CROWN
4-DR.
H/T
(WEST COAST)
$6774.

$6115.
(8492 BLT.)

HEADLIGHTS
CONCEALED
IN
new
GRILLE.
"Imperial"
NAME ON GRILLE,
and CREST on
HOOD, REAR DECK.

note
TRIPLE
OPENINGS FOR
SAFETY LT.

440 cid V8 (350 HP)

LE BARON
2-DR. H/T
(new)
$5898.
(4592 BLT.)

VINYL-COVERED
TOP
BEARS
TRADITIONAL
IMPERIAL
EAGLE
MEDALLION

BLOCK-
LETTER
"IMPERIAL" NAME IN CENTER OF REAR
BUMPER

(823
BLT.)

69

(INTRO.
9-19-68)

$5770.

CROWN 4 DR. H/T

LE BARON
4-DR. H/T
$6131. (14,821 BLT.)

TOTALLY
RESTYLED WITH
BULGE-SIDED "FUSELAGE" STYLING

225

22,089 JETS BLT. 1953

53 $1858. and up

JET

BY HUDSON

6 CYLS.

(1953–1954)

$1954

SUPER JET

DASH

IN ALL THE WORLD NO OTHER CAR LIKE THIS!

105" WB (THROUGH '54)

(ALSO 2-DR. JET FAMILY CLUB SEDAN and UTILITY SEDAN.)

$1621. UP

JET

$1885. (WEST COAST)

$1858.

54

GRILLE MODIFIED

4-DR. SED.

5.90 x 15

SUPER-JET

$1933.

ITALIA DASH

BUCKET SEATS

2-DR. CLUB SEDAN (4 DR. ALSO AVAIL.)

$4800.

ITALIA

JET-LINER

$2057.

note THAT EACH SERIES IS QUICKLY IDENTIFIED IN '54 BY AMOUNT OF CHROME SIDE TRIM.

4-DR. (2-DR. ALSO AVAIL.)

(ONLY 26 BUILT, ON SUPER-JET CHASSIS) 202 CID 6 (114 HP) 105" WB

SEE ALSO: son

TOTAL 1954 PRODUCTION = 14,224

1953 and 1954 ENGINE = 202 CID 6 (104 HP, or OPTIONAL 106/114 HP)

KAISER

(1946 – 1955)

KAISER-FRAZER CORPORATION

● WILLOW RUN, MICHIGAN

EARLY MODEL, BUILT 1946

EMBLEM

47

KAISER 6

K-100 or K-101 CUSTOM (112 HP)
(100 HP)

KAISER SPECIAL (65,062 BLT.)

with CORRUGATED BUMPER

SPECIAL = $1868.

CUSTOM = $2547.

AS IN FRAZER, 6-CYL., L-HEAD CONTINENTAL ENG. (ON ALL) 226.2 CID
6.50 × 15" TIRES
123½" W.B.

with PLAIN BUMPER   (CUSTOM 5412 BLT.)

ALL OVER THE MAP — YOU'LL FIND EXPERT KAISER AND FRAZER SERVICE

(EARLIEST '48s SOMETIMES CONSIDERED "1947½.")

48

K-481 or K-482 CUSTOM

$1967. $2557.

EARLY PRICES

new 7.10 × 15" TIRES

SEE ALSO: *FRAZER*

ILLUSTRATED AT CAPE COD, MASS.

SPECIAL = (90,588 BLT.) $2244.

CUSTOM = (1263 BLT.) $2466.

1948 PRICES

4-DOOR SEDANS ONLY

226.2 CID 6-CYL. CONTINENTAL ENG. CONTINUES (100 or 112 HP)

note 4 VERTICAL BUMPER GUARD ARRANGEMENT ('47½-'48 ONLY)

$3195.

new
4-DOOR
CONVERTIBLE
(54 BLT.)

new GRILLE

49-50

new 112 HP

TRAVELER
MODELS
FEATURE
FULL-OPENING
REAR
"HATCHBACK."

"TRAVELER"
MODEL NAME
IN SCRIPT

2-cars-in-one

$2088*

Kaiser Traveler

(22,000 BLT.)

(new)

$1995.
(SPECIAL)

$2195.
(DELUXE)

SPECIAL (29,000 BLT.)
DELUXE
(38,250 BLT.)   SEDAN

REAR 3/4 VIEW of
VIRGINIAN

new
GRILLE

new
VIRGINIAN

4-DOOR
HARDTOP
(946 BLT.)

GEAR RATIOS:
4.09; 3.91; 3.73
(OR 4.27 with OVERDRIVE)

new BODY TYPES
(ALL 4-DOOR
MODELS)

$2995.

1950 MODELS
REPLACED MARCH, 1950
BY TOTALLY RESTYLED
1951 MODELS.

Kaiser

$2275. new 2-door sedan DLX.

SEE ALSO: **"HENRY J"**

new HORIZONTAL BLADE GRILLE

K-511 = SPECIAL
K-512 = DE LUXE

1951
(TOTALLY RESTYLED)

115 HP @ 3650 RPM

SPECIAL = TRAVELER UTILITY SED. 2 DR. (1500 BLT.);
4 DR. SED. (43,500 BLT.); BUSINESS COUPE (1500 ");
2 DR. SED. (10,000); 4 DR. TRAVELER (2000); CLUB CPE.
(1500 BLT.) **DE LUXE** = 2 DR. " (1000 BLT.); 4 DR.
SEDAN (70,000 BLT.); 2 DR. SEDAN (11,000 "); TRAVELER
4 DR. UTILITY SEDAN (1000 BLT.); CL. CPE. (6000 BLT.)

new HIGH, ARCHED TOP, WITH HUGE WINDOW AREA!

DELUXE 4-DOOR
$2328.

new 118½" WB

The newest car in America!

Anatomic Design*

$1992.-$2433.
PRICE RANGE

HOOD ORNAMENT ADDED (ON ALL)

new "GOLDEN DRAGON" (with "ALLIGATOR" TYPE UPH., etc.)

HUBCAP VARIATION

Hydra-Matic
AUTO. TRANS. OPTIONAL
(THROUGH '55;
also OPT. ON 1951 FRAZER)

Built to Better the Best on the Road!

KAISER PRICED FROM $**2313.**

new "V" FIGURE ADDED to LOWER PART of FRONT and REAR EMBLEMS.

(1952 HAS NO CHROME FENDER FINS, HAS EXPOSED DECKLID HINGES.) LOWER BLADE IN GRILLE FORMS A BRIDGE BETWEEN THE BUMPER GUARDS. (K-521, 522) PLAIN HEADLIGHT RIMS on 1952. ('52)

1953 (new)

CAROLINA 2-DR. (K-538)

REAR DETAILS

DE LUXE TRAVELER (K-521)(K-531) IS DE LUXE SERIES)

$**1832.-2654.** 1952 PRICE RANGE

52-53

(K-530, 531, 532, 538) ('53)

118 HP '9 1/2" WB

$**2313.-3924.** 1953 PRICE RANGE

new 1-PC. WINDSHIELD

4 DR. $2241. ('52) $2619. ('53)

1952 : VIRGINIAN SPECIAL, VIRGINIAN DELUXE, DELUXE, and MANHATTAN SERIES.
1953 : CAROLINA, DELUXE, MANHATTAN, and DRAGON.

$**2654.** ('52)

1953 EXAMPLES SHOWN

SEDAN (15,450 BLT., '53)

MANHATTAN (K-522) '52 (K-532) '53

MERGES with WILLYS-OVERLAND, TO FORM KAISER-WILLYS.

(LENGTH EXAGGERATED) $**2650.** ('53)

(DRAGON IS K-530) $**3924.** (1277 BLT.) '53

KAISER-DARRIN DKF-161

OPTIONAL SUPERCHARGER GIVES 140 HP @ 3900 RPM (STD. HP 118)

'55

WITH (fiberglass body) (435 BLT.)

$**3668.** new LIGHTS and CONCAVE GRILLE

W. 6 CYL 161 CID WILLYS F-head ENGINE

('54)

'55 SIMILAR, BUT with HIGHER CHROME FIN TIP ON HOOD ORNAMENT (SEE ARROW)

7.10 x 15

54-55

1955 PRICES START AT $**2503.**

SIMILAR MODELS CONTINUED BY KAISER IN ARGENTINA (I.K.A.,) UNDER THE NAME of CARABELA. (1955 TO 1962.)

REAR ALSO RESTYLED

KING MIDGET

46-50

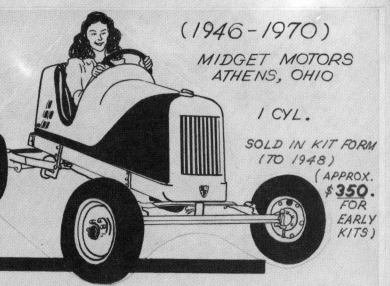

MIDGET MOTORS
ATHENS, OHIO

1 CYL.

SOLD IN KIT FORM
(TO 1948)

(APPROX. **$350.** FOR EARLY KITS)

76½" WHEELBASE (THROUGH '70)

AVAILABLE FULLY ASSEMBLED, 1949 ON

('55)

EARLY

51-57

(TOTALLY RESTYLED FOR 1951)

1 CYL., 7.3 TO 8.5 HP WISCONSIN ENGINE (TO 1966)

APPROXIMATELY 5000 BLT., 1946 ~ 1970

(RESTYLED EARLY '57)

LATER

57-70

30% HP INCREASE (TO 12 HP) FOR 1966

AFTER 1966, 12-HP KOHLER ENG.

DISCONTINUED 1970 **$1095.** IN '69

REPORTEDLY, NO MORE THAN 50 KURTIS CARS BLT.   FORD or CADILLAC V8 ENGINES USED. (6 CYL. STUDEBAKER ENG. IN EARLIEST MODEL)

FRANK KURTIS, FOUNDER
(ALSO BUILT RACE CARS)

(1949 - 1950)

CHOICE OF V-8 ENGINES 100" WB

KURTIS

KURTIS-KRAFT, INC.
LOS ANGELES and GLENDALE, CALIF.

KURTIS CONTINUED TO BUILD OTHER TYPES OF SPORTS and RACING CARS, AFTER EARL MUNTZ BEGAN PRODUCING "JET" *

* BECOMES MUNTZ JET IN '51, new 116" WB and ENLARGED TO 4-PASS.

1961 PRODUCTION: "170" (25,508 BLT.) "770" (49,268 BLT.)

LANCER

COMPACT   CHRYSLER CORPORATION

[DODGE]

Lancer 170 Two-Door Sedan

$2312.

770

WAGON

6 CYL. INCLINED O.H.V. ENGINE
170 CID with
101 HP @ 4400 RPM
or 148 HP @ 5200 RPM

new 61

INTERIOR

LARGER 225 CID 6 ALSO AVAIL.,
with 145 HP @ 4000 RPM OR 196 HP @ 5200 RPM

RWI-L , RWI-H
170   770

AIR COND., POWER STEERING and POWER BRAKES AVAIL.

1961

106½" WB

H/T
COMPACT DODGE LANCER

LANCERS BUILT 1961 and 1962 ONLY.
UNITIZED BODY SIMILAR TO PLYMOUTH VALIANT.

1962 MODELS : 170 = (2 DR.; 4-DR.; 4-DR. WAGON) (19,780 BLT.) $1951.-2306.
770 =   ("   "   "   ") (30,888 BLT.) $2052.-2408.
GT = H/T (13,683 BLT.) $2257.

LANCER 170 2-DOOR SEDAN 6 $1951.

$2256   (SLI-L) $2011. $2114.   ANCER 770 4-DOOR SEDAN 6

LANCER 170 4-DOOR SEDAN 6

WEST COAST

$2562.

$2306.

$2052.

LANCER 170 6-PASSENGER WAGON 6   LANCER 770 2-DOOR SEDAN 6

new GT

62

SLI

770
(SLI-H)

(SLI-P)

new GRILLE

DISCONTINUED AFTER 1962

1962

$2408. $2257.

232

LARK
BY STUDEBAKER
108½" WB
113", WAGONS

COMPACT SERIES
(1959 - 1963)

2-DR.
PLAY WAGON

$1925. ~ 2590.

59
new!

6 - CYL. OR V8 ENGINES
169.6 CID   259.2 CID
(90 HP)   (180 OR 195 HP)

note LOCATION of GRILLE MEDALLION on 1959 MODEL

$2296. (6)

$2431. (V8)

PROD.
98,744 6 CYL.
32,334 V8s

1959 LARK CARRIES "STUDEBAKER" NAME

REGAL H/T

LUXURY Reclining seats that let all the way down are an optional touch of sublime comfort. Seats are pleated, appointments tasteful. Colors are harmoniously keyed to exteriors.

H/T INTERIOR

DELUXE = 2 DR.; 4-DR.; 2-DR. WAGON; 4 DR. WAGON
REGAL = 4 DR.; H/T; CONVERTIBLE; "   "   (6 OR V8, EITHER SERIES)

60
GRILLE MEDALLION MOVED TO LOWER CENTER

4-DR. WAGON and CONVERT. ARE new IN 1960

REGAL CVT.

"LARK" NAME AT REAR END OF FRONT FENDER

"LOVE THAT LARK BY STUDEBAKER"

$1976. ~ 2756. PRICE RANGE

$2621. (6) OR $2756. (V8)

LARK

4 HEADLIGHTS ON new 113" WB CRUISER

6 has 112 HP

180 TO 225 HP

61

GRILLE EMBLEM MOVED; new PARK. LIGHTS; "LARK" NAME MOVED TO FORWARD END OF FR. FENDERS

"You have to drive The Lark to believe it!"

1961 PRODUCTION
6-CYL. = (41,035 BLT.)
V8 = (25,934 ")

$1935. ~ $2689. PRICE RANGE

$1935. ~ $2814. PRICE RANGE

PACE CAR AT 1962 INDY 500 RACE

DAYTONA CVT.

"LARK" IN CAPITAL LETTERS

DETAILS OF new GRILLE

new ROUND TAIL-LIGHTS

VIEWS OF DASH

62

6 CYL. = (54,397 BLT.)
V8 = (38,607 ")

SUNROOF OPTIONAL ON new
225 HP **DAYTONA**

$2308. UP

SLIDING REAR ROOF SECTION ON new REGAL LARK WAGONAIRE

STD. WAGONS $2430. UP

LARK NAME USED ONLY 1959-1963

63

new GRILLE AGAIN

$1935. ~ 2835. PRICE RANGE

REGAL LARK $2055. UP

SEE ALSO
Studebaker

REGAL WAGONS FROM **2550.**
DAYTONA " **$2700.**

74,201 TOTAL 1963 PRODUCTION

Lincoln

(SINCE 1921)
A DIVISION OF FORD MOTOR CO. SINCE 1922
(201 CONTINENTAL CVTS., 265 CLUB CPES.) **$4392.**

$4474   #56   #57
← CONTINENTAL (PACE CAR AT 1946 INDY 500 RACE)

#77 CLUB COUPE
$2318.

#76 CVT.

HEAVIER NAMEPLATE ON SIDES OF 1946 HOOD

$2883. **46** 66-H

292 CID V-12 ENG. (125 HP)

PRICE RANGE: **$2178.** TO **$4205.**

#73 SEDAN
$2337.

PUSHBUTTON DOORS ON ALL '46 MODELS

125" WB AS BEFORE, BUT "ZEPHYR" NAME DISCONTINUED.

RAISED HEXAGON AT CENTER OF 1946 HUB CAPS →

new LARGER BUMPERS

new HEAVIER GRILLE has BOTH HORIZ. and VERT. PCS.

(16,179 BLT.)

CUSTOM INTERIORS OPTIONAL AT EXTRA COST (THROUGH '48)

CONVENTIONAL DOOR HANDLES RETURN, ON STD. TYPES

$2554. ('48)

"Nothing could be finer "

"Lincoln" NAME IN CHROME ON SIDES OF HOOD and ON new PLAINER HUBCAPS

FINAL LINCOLNS with V-12 ENGINES (1948)

(19,891, 1947 PROD.; 6470, 1948 PROD.)

7-H   8-H
47-48

CONVERTIBLE

CONTINENTALS CONTINUE USE OF BUTTON DOOR OPENERS

1947 (738 BLT.); 1948 (452 BLT.)
CONTINENTAL $4746.

FINAL CONTINENTALS UNTIL 1956 MODEL

luxury car

CLUB COUPE
1947 (831 BLT.); 1948 (847 BLT.)

$4662.

LINCOLN

STD. 9EL SERIES (38,384 BLT.)
121" WB

$2527.

COUPE

new 2-PC. WINDSHIELD
(ON STANDARD
LINCOLNS ONLY)

1949

$2575.

COUPE

$3948.

new COSMOPOLITAN
125" WB (9 EH)

$3186.

49

(TOTALLY
RESTYLED)

(35,123 COSMOPOL.
BLT.)

BACK SEAT AREA (COSMO.)

The "custom touch" adds luxury to the 1949 Lincoln Cosmopolitan!

ALL
with new
V8 ENGINE
(L-HEAD)
336.7 C.I.D.
(152 HP)

LINCOLN

1949

$3238.

OEH
COSMOPOLITAN

SPORT SEDAN
(8341 BLT.)
$3240.

121" WB LINC.
"LIDO" CPE.
IS new
($2721.)

$2529.~3950. PRICE RANGE

new
GRILLE
IS
LOWER

50

SOME MODELS
PRICED
ONLY $2
HIGHER THAN
LAST
YEAR'S

$2505.~3891. PRICE RANGE

INTRO. DURING 1950,

final
LARGE-SIZED
COSMOPOLITAN

COSMO.
"CAPRI" CPE.
IS new

new
FULL-LENGTH CHROME MOULDING ALONG
BODY SIDES OF COSMOPOLITAN MODELS
(AND CONT'D.
ON STD.
LINCOLNS)

125"
WB

HP INCREASED
TO 154

121"
WB

51

LINCOLN SPORT SEDAN

new
GRILLE

COSMO.
SPORT SEDAN
$3182.

LINCOLN

new 123" WB $3198.

COSMOPOLITAN

H/T (5681 BLT.)
CAPRI

CVT. (1191 BLT.) $3665.

COSMO. IS NOW LOWER-PRICED SERIES, BELOW CAPRI.

52 (TOTALLY RESTYLED)

"Lincoln" NAME IN SCRIPT LETTERING, ABOVE new GRILLE

8.00 x 15" TIRES
TOTAL 1952 PRODUCTION, 31,992
new O.H.V. 317½ CID V8 (160 HP)

$3198.~3665. PRICE RANGE

$3226.   COSMOPOLITAN SEDAN

COSMOPOLITAN LETTERING DETAILS

53

(7560 BLT.)

HP INCREASED TO 205

H/T (12,916 BLT.)

CONV'T. DETAILS

CAPRI LETTERING DETAILS

CAPRI $3549.

new BLOCK "LINCOLN" LETTERING, ABOVE GRILLE WHICH NOW CONTAINS STYLIZED "V" and SMALL EMBLEM

TOTAL 1953 PRODUCTION, 41,962

$3226.~3699. PRICE RANGE

LINCOLN

H/T (14,003 BLT.)   CAPRI

(1951 BLT.) $4031.

CONV'T.

$3869.

new FENDER TRIM

54

new GRILLE

"LINCOLN" NAME NOW IN SCRIPT, and MOVED TO FRONT FENDER PANELS.

TOTAL 1954 PRODUCTION, 35,733

$3522.~4031. PRICE RANGE

$3563.~4072. PRICE RANGE

new 341 CID V8 (225 HP)

TOTAL 1955 PRODUCTION, 41,226

$3752.

55

CUSTOM IS LOWER-PRICED SERIES, PRICED from $3563.

SEDAN (10,724 BLT.)

CAPRI

H/T (11,462 BLT.)

$3910.

new GRILLE with ALL HORIZONTAL PIECES

new FENDER TRIM

new 126" WB, new 368 CID (285 HP)

56

new PREMIERE

new PANORAMIC WINDSHIELD

CAPRI H/T

FRENCHED HEADLIGHTS, and new PARK./DIRECTIONAL LIGHTS IN GRILLE

new CHROME SIDE SPEAR

ALSO, A REVIVED **Continental**

(SEE NEXT PAGE)

TOTAL 1956 PRODUCTION, 48,995

$4119.~4747. PRICE RANGE

LINCOLN

Continental

Mark II

Continental Division · Ford Motor Company

$ **9538.** ('56)

($157. MORE IN 1957.)

300 HP

126" WB

new CONTINENTAL STYLING DIFFERS FROM CAPRI, PREMIERE MODELS (THROUGH '60)

56-57

1ST CONTINENTAL SINCE 1948.

1957 PRODUCTION 37,870

$ **4649. ~ 5381.** PRICE RANGE

126" WB

NON-CONTINENTAL 1957 TYPES : *CAPRI* PRICED FROM $ **4649.**

(15,185 BLT.) COUPE (H/T)

PREMIERE (3676 BLT.) CVT.

$ **5149.**

new LANDAU 4-DR. H/T

$ **5381.**

(11,233 BLT.)
$ **5294.**

new 300 HP

57

LINCOLN

CAPRI new 131" WB (THROUGH '60) PREMIERE

H/T $4803.

CONTINENTAL MARK III

Unmistakably . . . the finest in the fine car field

58

(TOTALLY RESTYLED)

new 375 HP

CONT'L. HAS new CRISS-CROSS GRILLE PATTERN

CONT'L. NO LONGER HAS "SPARE TIRE BULGE" IN REAR DECK

9.00 × 14 TIRES

$6283. (CVT.)

TOTAL 1958 PRODUCTION, 25,871

430 C.I.D V8

$4902. ~ 10,230. PRICE RANGE

PREMIERE (CAPRI ALSO AVAIL.)

59

new GRILLE NOW ENCOMPASSES THE CANTED HEADLIGHTS

CUT TO 350 HP

CONTINENTAL MARK IV

$7056.

9.50 × 14 TIRES

TOTAL 1959 PRODUCTION, 30,375

LINCOLN

PREMIERE

$5698.

2-DR. H/T

4-DR.

430 CID V8
HORSEPOWER CUT TO 315 @ 4100 RPM
(new CARBURETOR)
LEAF SPRINGS REPLACE
COILS AT REAR

$5945.

new HOODED INSTRUMENTS

TYPICAL UPHOLSTERY (LEATHER and FABRICS)

60

DASH and INSIDE DOOR HANDLE

$6598.

CONTINENTAL MARK V

LANDAU 4-DR. H/T

2-DR. H/T

$5253. TO $10,230.
PRICE RANGE

FINAL YEAR FOR 2-DR. CONVERTIBLE

$6845.

"TOWN CAR" FORMAL SEDAN

$7056.

LIMOUSINE

$9208.

$10,230.

9.50 × 14 TIRES

1960 PRODUCTION = SEDAN (1093 BLT.); LANDAU 4-DR. H/T (4397 BLT.); H/T (1670 BLT.)
PREMIERE = SEDAN (1010 BLT.); LANDAU 4-DR. H/T (4200 BLT.); H/T (1365 BLT.)
CONTINENTAL MARK V = LIMOUSINE (34 BLT.); FORMAL SEDAN (136 BLT.); SEDAN (807 BLT.);
H/T (1461 BLT.); CONVERTIBLE (2044 BLT.); 4-DR. H/T (6604 BLT.)

LINCOLN CONTINENTAL

61

(TOTALLY RESTYLED)

new DASH

$ **6067.**

REDUCTION OF WHEELBASE TO 123" (THROUGH '63) and CUT IN H.P. TO 300 @ 4100 RPM (THROUGH '62)

DECK LID OPENS WHEN TOP MOVES

REAR DETAILS
9.00 x 14" TIRES

$ **6713.**

new Four-Door Convertible

ALL MODELS NOW KNOWN AS *Lincoln Continental*

ONLY 2 MODELS AVAIL.: 4-DOOR HARDTOP OR 4-DOOR CONVT.

(3212 BLT.)
6720.

62

new GRILLE and REAR END OF HARMONIZING DESIGN

(27,849 BLT.)
$ **6074.**

new THIN WHITE SIDEWALLS ON TIRES (9.00 x 14")

LINCOLN (CONTINENTAL)

63 new GRILLE

DETAIL OF CENTER-OPENING DOORS

$6916.

4-DR. CONV'T. with TOP UP
(3138 BLT.)

new 320 HP (THROUGH '65)

4-DR. H/T (28,095 BLT.)
$6270.

LIMOUSINE

4 DR. CVT. $6938. (3328 BLT.)

The luggage compartment is larger.

greater

4 DR. H/T
$6292.

64 (SLIGHTLY ENLARGED) new 126" WB (THROUGH '69)

interior spaciousness

4 DR. H/T (33,969 BLT.)

3" LONGER THAN BEFORE

4-DR. CONVERTIBLE CONTINUES THROUGH 1967

65

PRODUCTION :
4 DR. H/T (36,824 BLT.)
4-DR. CONVT. (3356 BLT.)

LIMO-ROOFLINE

DASH

CONV.
$6938.

$6292. →

320 HP

430 CID

new GRILLE with HORIZONTAL MOTIF

LINCOLN Continental
America's most distinguished motorcar.

HOOD ORNAMENT

126" WB
new 462 CID V8
(340 HP)
9.15 x 15
TIRES

$5485.
(15,766 BLT.)

$5750. 4-DR.

H/T (new)

↑ $6118. (WEST COAST)

$6383. (35,809 BLT.)
(WEST COAST)

66

$6383.
FRONT
END
DETAILS

4-DR. CVT.
$7016.
(WEST COAST)

Lincoln Continental for 1966:
unmistakably new, yet unmistakably Continental

(3180 BLT.)

(LEATHER UPH. STD. IN CVT., $111 EXTRA IN OTHERS)

FENDER-TIP LTS.
DISCONTINUED
UNTIL 1968

4 DR. CVT.
(2276 BLT.)

REAR DETAILS OF
1965 CONT'L.

$6449.

(INTRO. 9-30-66)

2-DR. H/T
(WITHOUT
VINYL TOP)
$6185.
WEST COAST

67 $5553.

GRILLE
SLIGHTLY
CHANGED,
and EMBLEM
REMOVED FROM SIDE
OF FRONT FENDER

(11,060 BLT.)

4-DR.
(WITH VINYL
TOP)

$5795.

(33,331 BLT.)

4-DR., STD. TOP
$6427. WEST COAST
(VINYL LANDAU
TOP $132. EXTRA)

1967 IS FINAL YEAR FOR THE
4-DOOR CONVERTIBLE.

DASH

new BROAD STEER. WH. HUB, DARKER-COLORED DASH ←

new REAR STYLING

68 (INTRO. 9-22-67)

Wraparound parking lights and taillights

new coupé roof line. ← (94/5 BLT.)

$5736.

$5970.

$6634. (WEST COAST)

DETAILS OF new GRILLE

RAISED HOOD ORNAMENT ELIMINATED (UNTIL 1972).

SEDAN (29,719 BLT.)

EMBLEM REPLACES NAME ABOVE GRILLE, DURING 1968

new SAFETY SIDE LIGHTS

(2 DR. H/T ALSO 9032 BLT. $5830.)

(29 258 BLT.) 4-DR.

2 SERIES NOW AVAIL.

69 new "COMPUTER DESIGNED" 460 CID V8 (365 HP)

(7770 BLT. '68) 23,088 BLT. '69) $6758.

$6063.

Continental

(INTRO. 9-27-68) new GRILLE WITH CONTINENTAL NAME ABOVE. (126" WB)

(NO 4-DR. MK. TYPES UNTIL 1980 MK. VI)

117.2" WB

Continental Mark III.

MK. III INTRO. 4-5-68

(NEW)

MERCURY

new SPORTSMAN CONV'T. $2209.

(WOODEN BODY PANELS)

2-DR. SED. (13,108 BLT.)

(205 BLT.)

46 new GRILLE

NEW INTERIORS

WAGON (2797 BLT.)

$1448.

RARE CANADIAN **MONARCH** VARIATION OF 1946-1948, A MERCURY-SIZED CAR SOLD BY CANADA'S FORD DEALERS. (SEE BACK PGS. FOR MORE!)

"COUPE-SEDAN" (24,163 BLT.) $1495.

$1711. ALL-STEEL CONVERTIBLE

new GRILLE

(6044 BLT.)

SLOGAN: "STEP OUT WITH MERCURY"

$1509.

2-DR. SEDAN $1592. (34 BLT. 1947)

STATION WAGON (3558 BLT., 1947; 1889 BLT., 1948)

new BUMPERS

"MERCURY" NAMEPLATE ABOVE GRILLE NOW HAS UNDERLINED BLACK BACKGROUND INSTEAD OF RED AS IN '46.

47-48

BORDER OF GRILLE IS NOW CHROME-PLATED.

CHROME

$2207.

$1660.

1947 INTERIORS

More OF EVERYTHING YOU WANT WITH *Mercury*

MERCURY

$2716. (8044 BLT.) new 2-DR. WAGON

CONVERTIBLE (16,765 BLT.) $2410.

CLUB CPE.

49 TOTALLY RESTYLED

(120,6/6 BLT.)

7.10 × 15" TIRES

new 255.4 CID V8 (new 110 HP)

SPORT SEDAN (155,882 BLT.)

Make your next car **Mercury**

BRIER INN

1949

FROM $1997. (CLUB CPE.)

118" WB

$2031.

(TOTAL 1950 PRODUCTION = 334,081)

new EMBLEMS AT EITHER END

"Better than ever"

50

PACE CAR AT 1950 INDY 500 RACE

$2412. CVT. (8341 BLT.)

1950

LARGE PARK. LIGHTS AT ENDS of GRILLE

CLUB COUPE $1980.

$1947. ~ 2530. PRICE RANGE

4-DR. SPORT SEDAN (157,648 BLT.) $2000.

"MONTEREY" COUPE
CLUB COUPE (ALSO AVAIL. SINCE '50, w. CANVAS or VINYL TOP COVERING)

$1947.

new GRILLE and new EMBLEMS

Nothing like it on the **Road!**

1951

new VERT. TAIL- LIGHTS

51

new OPTIONAL **MERC-O-MATIC** AUTO. TRANS.

AUTO. TRANS.

LARGER BACKLIGHT (REAR WINDOW)

for "the buy *of your life!"*

new 112 HP

MERCURY Merc-O-Matic Drive...or B-W Overdrive

CUSTOM

NEW 125 H.P. HIGH-COMPRESSION V-8

MONTEREY $2225. H/T

hardtop

H/T (24,453 BLT.)

new DASH

255.4 CID

52 (TOTALLY RESTYLED)

118" WB

new HOOD SCOOP

FROM $1987.

new BUMPER-GRILLE

MONTEREY SEDAN

$2115.

7.10 x 15" TIRES, (7.60 x 15" ON '52 CVT. AND '53 "MONTEREY" MODELS)

GET THE FACTS — AND YOU'LL GO FOR THE NEW 1953 **MERCURY**

LOWER-PRICED "CUSTOM" SERIES IN $2004.~2117. PRICE RANGE

MONTEREY

$2133.

DASH

new DECK-LID MEDALLION

POWER STEERING

H/T (76,119 BLT.) $2244.

new HORIZ. REAR FENDER TRIM

H/T

POWER BRAKES

3 new POWER OPTION CHOICES

53

new GRILLE 118" WB

4-WAY POWER SEAT

MERCURY

CUSTOM $2251.

SEDANS

8-PASS. WAGON (11,656 BLT.) $2776.

MONTEREY

new 161-horsepower engine

DASH

TOTAL 1954 PRODUCTION = 256,729

$2581. →

"SUN VALLEY" (new)

THE CAR THAT MAKES ANY DRIVING EASY

(9761 BLT.)

H/T (79,533 BLT.) $2452.

54

new GRILLE

55

TOP-LINE MONTCLAIR SERIES

19 new

new 119" WB

new DASH

TOTAL 1955 PRODUCTION 434,911

new 292 cid O.H.V. V8 ENG. 188 HP (or 198 HP in MONTCLAIR)

$2686.

CUSTOM

$2277.

new REAR STYLING →

$2465.

(69,093 BLT.)

NEW, EXCLUSIVE POWER LUBRICATION.

$2844.

(11,968 BLT.)

MONTEREY

$2400.

(70,392 BLT.)

new PANORAMIC WINDSHIELD

new SIDE TRIM

7.10 x 15" TIRES

$2631.

$2712.

FROM $2218. (CUSTOM 2-DR.)

new GRILLE and HOODED HEADLIGHTS

(71,588 BLT.)

MONTCLAIR

SUN VALLEY

(1787 BLT.)

MERCURY

MEDALIST
(new) (1956 ONLY)
(LOWEST-PRICED MODEL)

56

$2254.

(20,854)

MONTEREY

CUSTOM

119" WB

VOYAGER
(IN Montclair SERIES)

118" WB
ON WAGONS

"PHAETON"
4-DR. HARDTOP
(23,493 BLT.)

$2835.

REAR 3/4 DETAIL

MONTCLAIR

INTERIOR

CONVERTIBLE
(7762 BLT.)
$2900.

For 1956 — the big move is to
THE BIG MERCURY

new
312 CID V8
(210, 225 OR
235 HP)

MONTCLAIR
H/T (50,562 BLT.)
$2765.

CLOSE
VIEW
OF
new
GRILLE

MERCURY

BIG M | for '57

HIGH BEAM
LOW or HIGH BEAM

QUADRI-BEAM HEADLAMPS
(LATER MODELS)

MONTEREY

(EARLY) MONTCLAIR

PACE CAR AT 1957 INDY 500 RACE

$3758.

57 with DREAM-CAR DESIGN (TOTALLY RESTYLED)

(LATER)

new 122" WB 255 HP

FRONT ROOF VENTS on TURNPIKE CR. (290 HP)

CONVENTIONAL STATION WAGON | NEW BIG M STATION WAGON

MERCURY ELIMINATES THE LIFT GATE, LOWERS THE TAIL GATE

new TURNPIKE CRUISER

THERE'S ONLY ONE SIDE PILLAR IN THE NEW MERCURY COMMUTER

VOYAGER | 2 and 4-DR. WAGONS

THE OPEN-AIR FEELING OF A HARDTOP

ONLY 2 HEADLIGHTS on EARLY MODELS

COLONY PARK

$3677.

6 wagons

BIG *new* WEDGE TAIL-LIGHTS

CENTER OF BACKLIGHT OPENS, on TURNPIKE CR.

5-7. NEW MONITOR CONTROL PANEL, TACHOMETER, AVERAGE SPEED COMPUTER

2 V8 ENGINES AVAIL. 312 CID (255 HP); 368 CID (290 HP)
1957 MODELS: MONTEREY = PHAETON 4-DR H/T (22,475 BLT.); SEDAN (53,839 BLT.); PHAETON 2-DR H/T (42,199 BLT.); 2-DR. (33,982 BLT.); PHAETON CONVT. CPE. (5033 BLT.) **MONTCLAIR** = PHAETON 4-DR. H/T (21,567 BLT.); SEDAN (19,836 BLT.); PHAETON 2-DR. H/T (30,111 BLT.); PHAETON CVT. CPE. (4248 BLT.)
TURNPIKE CRUISER = H/T (7291 BLT.); 4-DR. H/T (8305 BLT.); CONVT. (1265 BLT.) **WAGONS** = COMMUTER 2-DR. (4885 BLT.); VOYAGER 2-DR. (2283 BLT.); COMMUTER 4-DR. (11,990 BLT.); COLONY PARK, 4 DR., 9-PASS. (7386 BLT.); COMMUTER, 4 DR., 9-PASS. (5752); VOYAGER, 4-DR., 9-PASS. (3716 BLT.)

MERCURY

PRICED FROM $2547. TO $4118.

THE ALL-NEW PARK LANE

(LOW PRICED MEDALIST RETURNS FOR 1 MORE YEAR.)

$3944.

WHEEL COVER

58

122" WB (126" ON PARK LANE)
PARK LANE REPLACES TURNPIKE CRUISER

ENGINES :
312 cid V8
(235 HP)
383 cid V8
(312 OR 330 HP)
new 430 cid V8
(360 HP)

TOTAL 1959 PRODUCTION = 156,756

$2768. ~ 4206. PRICE RANGE
(MONTEREY 2-DR.) (PARK LANE CVT.)

ENGINES :
312 cid V8
(210 OR 280 HP)
383 cid V8
(280 OR 322 HP)
430 cid V8
(345 HP)

20ᵗʰ ANNIVERSARY

'59 MERCURY

MONTEREY 4-DR. H/T
(11,355 BLT.)
$2918.

"BUILT TO LEAD
— BUILT TO LAST "

59. ENGINE

new GRILLE VARIES IN APPEARANCE, DEPENDING ON ANGLE FROM WHICH IT IS VIEWED (SEE ALSO NEXT PG.)

new ENLARGED WINDSHIELD AREA

(CONT'D.)

MERCURY

MONTEREY SEDAN

$3357.

(7375 BLT.)

MONTCLAIR

(28,892 BLT.)
$2721.

59
(CONT'D.)

1959

FANCIER REAR STYLING
ON MONTCLAIR

(7206 BLT.)
$4031.

(15,122 BLT.)
COMMUTER 4-DR.
6-PASS. WAGON

$3215.

VOYAGER 4-DR., 6-PASS.
(2496 BLT.) WAGON

$3793.

VOYAGER

PARK LANE
4-DR. H/T
CRUISER
(ABOVE)
has SPECIAL
REAR SIDE
TRIM

COMMUTER

WHEEL
COVER

126" WB
(128"
ON
Park Lane)

COLONY PARK
$3932.
(6-PASS.)

WAGON DETAILS

SLIP THE THIRD SEAT UNDER THE FLOOR

$3330. (9-PASS.)

COMMUTER

MERCURY

$2631.

2-DR. H/T (MONTEREY)

MONTCLAIR

2 DR. MONTEREY

(21,557 BLT.)

CVT.

(6062 BLT.)

$3077.

(RESTYLED) **60**

4-DR. H/T

(MONTEREY)

9-PASS. COMMUTER

$3240.

COLONY PARK 9-PASS.

$3950.

(7411 BLT.)

$3858.

PARK LANE 4-DR. H/T "CRUISER" (5788 BLT.)

126" WB ON ALL MERCURYS (1960 ONLY)

$4018.

PARK LANE CVT.

(1525 BLT.)

5 VERTICAL CHROME PCS. IDENTIFY PARK LANE. COLONY PARK has 6 PCS, and MONTCLAIR has 3.

(ONLY 2 STATION WAGON MODELS AVAIL. 1960, EACH ILLUSTRATED ON THIS PAGE.)

ENGINES: 312 cid V8 (205 HP)
383 cid V8 (280 HP) OR
430 cid V8 (310 HP)

1960 PRICE RANGE = $2631. ~ 4018.

TOTAL 1960 PRODUCTION = 161,787

254

MERCURY

METEOR 600
new series
V8 or new 6

the better low-price cars

METEOR 800

H/T
$2774.

MODEL NAME AT FRONT END OF DOOR →

Meteor 800

61

4-DR.
H/T
(9252 BLT.)
MONTEREY

MONTEREY

$2878.

COLONY PARK

$2924.
COMMUTER

(7887)
BLT.

$3120.
UP

MODEL SERIES: METEOR 600 = $2535. 2589. (18,117 BLT.)
METEOR 800 = $2713. 2839. MONTEREY = $2871. 3128.
4 DR. STATION WAGONS = $2924. ~ 3191.   4 DR. (22,881 BLT.)
 COMMUTER (8945 BLT.)   H/T (10,942 BLT.)
 COLONY PARK   }(7887 BLT.)   4 DR. H/T (9252 BLT.)
  "   "   (9-PASS.)   CONVERTIBLE (7053 BLT.)
 COMMUTER (" ") (6 BLT.)

125,792
TOTAL 1961
PRODUCTION

$2278. ~ 3738.
PRICE RANGE

METEOR
116½" WB

62

S-33
DASH

TOTAL 1962 PRODUCTION = 190,560   METEOR

MONTEREY
CUSTOM

new GRILLES

S-33
WHEEL COVER

(18,975 BLT.)

MONTEREY

new
TAIL-LIGHTS
AT TOP OF
FENDERS

6 or V8 (101 to 330 HP)

BA 953G

2672.

(S-55 MODELS ALSO)

"BEST-LOOKING BUYS...NOW IN EACH SIZE"

MERCURY

(4865 BLT.) **$2628.** **METEOR** **$2278.** and up

METEOR S-33

FINAL METEOR. 6-CYL. MERC. ENG. ONLY IN COMET AFTER '63.

MONTEREY
$2930.

H/T (3879 BLT.)

new OPENING "BREEZEWAY" BACKLIGHT →

$2995. 4-DR H/T

(1692 BLT.)

CONSOLE (S-55)

63

S-55 CONVERTIBLE and INTERIOR ←→

MONTEREY CUSTOM

(1379 BLT.)
$3900.

ENGINES:
170 CID 6 (101 HP)
221 CID V8 (145 HP)
260 CID V8 (164 HP)
390 CID V8 (250, 300 OR 330 HP)
406 CID V8 (385 OR 405 HP)

TOTAL 1963 PRODUCTION = 141,392

$2719.

(7298 BLT.)

$3083.

MONTRY. CUSTOM MARAUDER ('63)

METEOR CUSTOM (3636 BLT.)

COUNTRY CRUISER

(13,936 BLT.)

COLONY PARK and INTERIOR →

256

MERCURY

V8s ONLY $3236. UP — 2-DR. H/T

COMMUTER WAGON — Commuter station wagon

$3434. UP COLONY PARK

No finer car in the medium-price field

MONTEREY
FROM $3202. WEST COAST 120" WB (ALL MOD.)

250 HP V8

64

4-DR. MARAUDER H/T $3413.
$3567. WEST COAST

MONTCLAIR

MARAUDER SO-CALLED "FASTBACK"-STYLE H/T (6459 BLT.)
$3127.

$3127. 2-DR. H/T (BREEZEWAY ROOFLINE) (2329 BLT.)

CLOSE-UP OF DOOR (PARK LANE) SHOWN ABOVE

WEST COAST $3799. EASTERN $3413.
4-DR. MARAUDER H/T

PARK LANE (300 HP) INTERIOR →

$3359.
2-DR. H/T (BREEZEWAY ROOFLINE) (1786 BLT.)

DASH

1964 ENGINES: 390 CID V8 (250, 266, 300 or 330 HP)
427 CID V8 (410 or 425 HP)

1964 PRICE RANGE = $2819. ~ 3549.

NOTE THE DIFFERING ROOFLINE STYLES, AS ILLUSTRATED

MERCURY

MONTEREY, MONTCLAIR, PARK LANE and WAGON MODELS

$2767.~3599.
PRICE RANGE

NOW IN THE LINCOLN CONTINENTAL TRADITION

TOTALLY RESTYLED

65

1966 MERCURYS ON NEXT PAGE

YEARLY PRODUCTION:	
1954 = 13,095	1955 = 6096
1956 = 9068	1957 = 15,317
1958 = 13,128	1959 22,309
1960 = 13,103	1961 853
1962 = 412	

(1961 and 1962 MODELS WERE ACTUALLY 1960 LEFTOVERS.)

AMERICAN MOTORS IMPORTED

Metropolitan

(1954-1961)

542 COUPE $1445.
('54~'56)

85" WB (ON ALL)

$1469.
('54~'56)

"HOOD SCOOP"

ASSEMBLED in ENGLAND BY AUSTIN, FOR U.S. MARKET

541 CVT.

4 CYL., O.H.V. 73.8 CID AUSTIN ENGINE (42 HP)

54-55

SERIES 54   (ALSO EARLY '56)

TYPES CONT. TO EARLY 1956

note:
VENT WINDOWS, OUTSIDE TRUNK DOOR ADDED DURING 1959 MODEL YEAR

1960 MODELS ILLUSTRATED BELOW

56-61

SERIES 56 "1500"
new GRILLE and SIDE TRIM

561 CONVERTIBLE
$1697. ('60~62)

52 HP
90.9 CID 4 CYL.

562 COUPE (4 PASS.)
$1673. ('60~62)

AT YOUR
RAMBLER-METROPOLITAN DEALER

* = EARLY 1962 ALSO

MERCURY CONT'D. ON NEXT PAGE

Mercury

390 CID V8 (265 or 275 HP)
410 CID V8 (330 HP)
428 CID V8 (345 HP)

COLONY PARK WAGON

Dual-Action Tailgate. Swings down like a regular tailgate for cargo. Or swings aside like a door for people.

9"
3
ONS)

walnut-toned paneling.

-SEAT = $3893.; 3-9EAT = $3988.
UNGRAINED COMMUTER
WAGONS
ALSO
VAIL.)

MONTEREY, MONTCLAIR, PARK LANE, S-55 MODELS

66

123" WB

MONTEREY

new BROUGHAM and MARQUIS MODELS

new SIDE TRIM on COLONY PK. WAGON
(18,690 BLT.)

67

20. 9-30-66)

$3657. UP

new PROTRUDING CENTER SECTION OF FRONT END

STD.
390 CID V8
has new 270 HP
(S-55 DISCONTINUED)

ENGINES:
390 CID V8
(265, 280 or 315 HP)
428 CID V8
(340 HP)

(INTRO. 9-22-67)

new FRONT CORNERING and REAR SIDE LTS.

new SWEPT-BACK ROOFLINE

FRONT DETAIL

68

First hardtop with yacht-deck vinyl paneling.

PARK LANE
$3575.

BROUGHAM
4 DR. H/T
BROUGHAM OPTION $272. EXTRA

new GRILLE

1968

Mercury

new 124" WB
(121" ON MARAUDER, WAGONS)

STD. 390 CID V8
(265/280 HP)

(25,604 BLT.) **COLONY PARK** $3895

(new MONTEREY, MONTEREY CUSTOM WAGONS ALSO AVAIL.)

$3919.

the new **MARQUIS.** (now INCLUDES ADDITIONAL BODY TYPES

MARQUIS
429 CID V8
(320 HP)

(9907 BLT.)
2-DR. H/T
$4098.

69

(INTRO. 9-27-68)

(new MARQUIS BROUGHAM ALSO AVAIL.)

new CONCEALED HEADLIGHTS ON MARQUIS, MARAUDER

8.25/8.55 x 15 TIRES

$4091.

note OWN REAR STYLING ON new MARAUDER (X-100)
2-DR. H/T
429 CID V8
(360 HP)
(5635 BLT.)

Lincoln-Mercury leads the way

$4270.

(MONTEREY and MTY. CUSTOM have ONE-TIER HORIZ. GRILLE and 4 EXPOSED HDLTS.)

PRODUCTION

1966 = 343,149 BLT., (INCL. COMET)	(MERCURY ONLY, 153,680)	$2783.~3614. PRICE RANGE
1967 = 354,923 "	(" " , COUGAR)	96,309	2904.~3986.
1968 = 360,467 "	" " " , MONTEGO)	142,048	3052.~3888.
1969 = 398,262 "	" " " , MONTEGO)	182,497	3158.~4262.

260

MUSTANG

ROY C. McCARTHY,
MUSTANG ENGINEERING CO.,
SEATTLE AND RENTON, WASH.

(1947-1949)

4- CYL. HERCULES ENGINE
59 H.P.   65 M.P.H.
NO DEALERSHIPS; FACTORY ORDERS ONLY

49

ALUMINUM BODY
102" W.B.
5.50 × 15" TIRES

TUBULAR CHASSIS FRAME

| THIS MUSTANG NOT AFFILIATED WITH FORD MOTOR COMPANY. |

$1235.

MUSTANG *Ford*   BY **Ford Motor Co.**

(STARTS APRIL, 1964)

6 OR V8
(170 CID) (260 CID)

65 *new!*

standard-equipment

STD. TYPE w/o
GRILLE LIGHTS →

(bucket seats, full carpeting, vinyl interior,
floor-mounted transmission)

CVT. IS PACE CAR AT
1964 INDY 500
RACE.

WHEEL
COVER

Surprisingly spacious trunk

REAR

STD. DASH

ALL CIRCULAR GAUGES ON
DE LUXE DASH
(BELOW)

$2368 *f.o.b.* Detroit
— AND UP —

INTRO.
PRICE

New luxury instrument panel

options *INCLUDE*:

a 289 cu. in. V-8. Four-on-the-floor. Tachometer and clock
combo. Special handling package. Front disc brakes—

STANDARD
SIDE
EMBLEM

NOTE
MESH
GRILLE
ON '65.

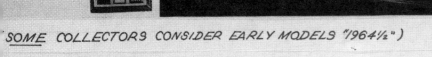
SOME COLLECTORS CONSIDER EARLY MODELS "1964½")

(CONT'D. NEXT PAGE)

MUSTANG

Unique Ford GT stripe — badge of America's greatest total performance cars!

"2 + 2" BACK SEAT →

New integral arm rests — courtesy lights

INTERIOR VIEWS (ABOVE)
$2372.

"2 + 2" FASTBACK $2589.

NOTCHBACK HARDTOP $2372.

Mustang GT

CONVERTIBLE $2614.

65 (CONT'D.)

EXTRA (FOG) LIGHTS IN GT GRILLE

IDENTIFYING RACING STRIPES ON GT →

MUSTANG

1965 PRODUCTION (SINCE APRIL, 1964)
#07 2-DR. H/T (NOTCHBACK) (501,965 BLT.)
#08 CONVERTIBLE (101,945 BLT.) = $2614. UP
#09 FASTBACK COUPE (77,079 BLT.)

(303,408 MUSTANGS BLT. BEFORE 1965 CALENDAR YEAR)

ENGINES: 170 CID 6 (101 HP) (TO 9-25-64)
200 CID 6 (120 HP) (9-25-64 ON)
260 CID V8 (164 HP) (TO 9-25-64)
289 CID V8 (200 HP) (9-25-64 ON)
(225 OR 271 HP OPTIONAL ON ALL 289 CID V8s)
(VINYL-COVERED ROOF AVAIL.)
(2 DR. H/T)

Mustang

FENDER BADGE ON GT

108" WB
200 CID 6 (120 HP)
289 CID V8 (200 HP)

SIDE EMBLEM: CHROMED HORSE WITH RED, WHITE and BLUE VERTICAL BARS

MUSTANG! MUSTANG! MUSTANG!
(SLOGAN)

REAR

6.95 × 14 TIRES

1966 PRODUCTION : 607,568

2 + 2
$2924.
(35,698 BLT.)

66

MESH IN GRILLE REPLACED BY HORIZONTAL STRIPS FOR 1966.

DASH

Mustang's new instrument panel groups five easy-to-read dials

$2970.

CVT.
(72,119 BLT.)

VARIATION →

$2416.

H/T (499,751 BLT.)

$2734.
(WEST COAST)

$2416. ~ 2653. PRICE RANGE
H/T   CVT.

ENGINES : 200 CID 6 (120 HP)
289 CID V8 (200, 225 OR 271 HP)
new 390 CID V8 (320 HP)

DASH

new GRILLE

67

FROM $2791. (WEST COAST)
H/T (356,271 BLT.)

TYPE WITH PLAIN REAR PANEL
2 + 2

$2921.
(WEST COAST)
SPORTS SPRINT

H/T

SLOGAN :
Take the Mustang Pledge.

GT 2 + 2
$3242.
(WEST COAST)

1967 PRODUCTION : 472,121

CALENDAR YEAR : 580,767

MUSTANG

Carroll Shelby Presents _The Road Cars_...
G.T. 350 and G.T. 500 for 1967

500

SHELBY G.T. **67** MODIFIED
350 (SHELBY TYPES)

Shelby Cobra

SHELBY COBRA DASH

68 new GRILLES

6.95 x 14 TIRES

new SIDE TRIM

$2712. UP (STD. 6)

GT 2+2

STD. 2-DR. H/T PRICED

FROM $2602.

STD. 6 and V8 REDUCED 5 HP

new 427 V8 AVAIL. (390 HORSEPOWER)
new 302 V8 ALSO (230 HP)

1968 PROD.: 317,404

CONV'T. (25,376 BLT.) $2814.
CALENDAR YEAR PROD.: 345,194

MUSTANG
FORD DIVISION Ford

69 *new* GRILLE

GT H/T $3129.

Ford's Exclusive "Shaker" scoop actually protrudes through the hood—rams air directly into the carburetor under full throttle.

MACH 1

$3480.
1969

MACH 1
SPORTSROOF

MUSTANG "E" SPORTSROOF $3078 WITH 250 CID 6, 155 HP)

MACH 1 ENGINE: 351 CID V8 (250 HP)

GT SPORTSROOF (FASTBACK) →

GRANDE

$3329 GRANDE *has* LOW REAR-FACING "SCOOP" LIKE GT MODELS.

STD. ENGINES: 200 CID 6 (115 HP) 302 CID V8 (220 HP)

SHELBY CARS STILL CUSTOM-CRAFTED WITH FORD PARTS, BUT NO LONGER BEAR A CLOSE RESEMBLANCE TO MUSTANG.

GT CVT. $3343. (WEST COAST)

(WEST COAST PRICES IN SMALLER PRINT, ABOVE.)

STANDARD PRICES, PRODUCTION FIGURES FOR 1969: H/T (128,458 BLT.) $2635.
FASTBACK (60,046 BLT.) $2635.; BOSS 302 V8 FASTBACK (1934 BLT.); CVT. (14,746 BLT.) $2849.
GRANDE H/T (22,182 BLT.) $2866. $3588.
MACH I FASTBACK
(40,970 BLT.) $3271.

NASH

(1917~1957)

FIRST 2 DIGITS IN MODEL NUMBER ARE THE YEAR MODEL

NOW 6-CYL. ONLY
(THROUGH '54)

600 DLX.

112" WB
L-HEAD ENG.
$1342.

121" WB
112 HP OHV

4663

$1453.

new MEDALLION and PK. LITES

46 PROD.

AMBASSADOR

6148 (LATER '45)
98,769 (DURING '46)

$1929.

4640

MODEL 4664 AMB.
SUBURBAN SEDAN with
WOODEN PANELING
(272 BLT. '46; 595 BLT. '47)

new GRILLE

AMBASS. SEDAN
IS PACE CAR
AT 1947
INDY 500
RACE

EL SEGUNDO, CALIF. and TORONTO, ONT. BRANCH PLANTS
PURCHASED THE PRECEDING YR.
MEXICO CITY
PLANT OPENS
6-18-47.

4748 (fastback)
4740 (bustle back)

4740
(600)
4760
(AMB.)
$1809.

"You'll be Ahead with
Nash".

47

new CHROMED EXTENSIONS at EITHER SIDE
of UPPER GRILLE PORTION

PROD.:
113,315

600

4842

$1874.

48

EXCEPT ON "600"
new HIGHER BELT LINE
CHROME FOR
1948
$2345.

COUPE
$1478. (BROUGHAM)

AMBASSADOR
SUPER

"FASTBACK"
SEDAN

4868

new CVT. (1,000 BLT.)
AMBASSADOR

4871

"BUSTLE
BACK"
SEDAN

600

4840

DASH
(MORE
DETAIL
NEX
PA

$1587.

AMB. 4863
or 4843 600

(COUPE)

TO $2047.

TO $2047. (AMB.
CUSTOM)
4873

600 = DELUXE BUSINESS CPE.; SUPER TRUNK-BACK 4-DR.; 2-DR. BROUGHAM; FASTBACK 4-DR.
CUSTOM TRUNK-BACK 4-DR.; 2-DR. BROUGHAM; FASTBACK 4-DR. AMBASSADOR = SIMILAR SUPER and
CUSTOM MODELS, ALSO SUPER SUBURBAN (130 BLT.) and CUSTOM CONVT. (1000 BLT.)
(CONT'D. NEXT PAGE)

NASH

FULL VIEW OF INTERIOR

48
(CONT'D.)

118,621 BLT.

You'll be Ahead with *Nash*

Great Cars Since 1902

$1916.

"SUPER" and "CUSTOM" are new

AMB. SUPER (MODEL NAME ON SIDE OF HOOD.)

4860

600 82 HP SUPER -DR. 1949

has "600" in CHROME, ON FRONT FENDER PANEL.

$1786.

EL SEGUNDO, CALIF. PLANT OPENS 10-48

49
TOTALLY RESTYLED new *Airflyte* MODELS
(NO CVTS.)

PHANTOM VIEW

ONE SINGLE WELDED UNIT!

with Girder-built Unitized Body and Frame
...Airliner-styled interiors...
Cockpit Control...Uniscope...
Matched Coil Springs on all
Four Wheels...Twin Beds...
Uniflo-Jet Carb

AMBASSADOR 112 HP
$2195. UP

1949 PRODUCTION: 600 OR **AMBASSADOR**, AVAIL. IN
142,592 SPECIAL, SUPER SPECIAL OR CUSTOM MODELS
ENGINES =
172.6 CID 6 (82 HP) "600" $1786. ~ 2363. PRICE RANGE
234.8 CID 6 (112 HP) AMBASSADOR

NASH

BACKLIGHTS ENLARGED

WITH *HYDRA-MATIC DRIVE*

5078

$2223.

191,865 BLT.

The Ambassador Custom

115 HP

50

...NEW SUPER-POWER ENGINES!

The Statesman 85 HP

(REPLACES 600)

new **SLIDING GLOVE DRAWER** THICKER BUMPER GUARDS

new **Rambler** *also avail.* AT NASH DEALERS

(SEE **RAMBLER** SECTION)

$1633. ~ 2223. PRICE RANGE

2 DR. BROUGHAMS, 2 DR. SEDANS and '4 DR. SEDANS, IN SUPER, SUPER SPECIAL, CUSTOM MODELS (STATESMAN or AMBASSADOR)

Airflytes for 1951

SUPER SEDAN 5148

$1955.

STATESMAN

$2099.

5159 CUSTOM 2-DR.

51

new **GRILLE** with ALL-VERTICAL PIECES

note PROTRUDING BUMPER GUARDS ON AMBASSADOR's

— new BUMPERS.

new **GRILLE** with VERTICAL PCS.

New sky-flow fenders

TRUNK DETAILS

$2304.

5169

AMBASSADOR SUPER

$2330.

5168

RECLINING SEATS (with BODY CENTERPOST NOT SHOWN, IN ORDER THAT SEAT DETAIL CAN BE SEEN.)

new **PARKING LIGHTS**

103,585 BLT.

NASH
(TOTALLY RESTYLED FOR 1952)

Golden Airflytes 50TH ANNIVERSARY (OF RAMBLERS)

Pinin Farina, STYLIST

$2332. 5255 ('52)

new 88 HP

new 114¼" WB
STATESMAN CUSTOM

AMBASSADOR CUSTOM ← 120 HP

5275
$2716.

new 121¼" WB

7.10 × 15

52-53

'53 has CHROMED HORIZ. SPACERS ON COWL AIR SCOOP) $2332.

152,141 BLT.*   153,753 BLT.*   *= INCL. RMBLR.

5355

('53) 100 HP

'53 with new STRIPS OF CHROME ON VENT

O.H.V. AMBASSADOR ENG.

DASH ('53)

('53)
AMBASSADOR COUNTRY CLUB
$2829.   5377

Ambassador Country Club

MODELS: **STATESMAN** = #5445 SUPER 4-DR. SEDAN; #5446 SUPER 2-DR. SEDAN; #5455 CUSTOM 4-DR. SEDAN; #5457 CUSTOM COUNTRY CLUB H/T

AMBASSADOR = 5475 " " " " ; 5477 " " "
#5465 SUPER 4-DR. SEDAN; #5466 SUPER 2-DR. SEDAN

$2110. ~ 2735. PRICE RANGE

$2417.

STATESMAN SUPER

$2110.
5446

67,192 BLT.

54

GRILLE MODIFIED (NOW CONCAVE)

5475

AMBASSADOR SUPER

5465

new BORDERS AROUND MODIFIED GRILLE

ST. OR AM. CUST. MODELS have REAR-MOUNTED "CONTINENTAL" SPARE TIRE.

AMERICAN MOTORS CORP. FORMED BY MAY 1, 1954 NASH-HUDSON MERGER.

AMBASSADOR CUSTOM

$2600.

STATESMAN = 195.6 CID 6 (110 HP)   AMBASSADOR = 252.6 CID 6 (130 OR 140 HP)

'55 NASH
(6 or V8)

STATESMAN 6 = $2215.~2495.
AMBASSADOR 6 = $2480.~2795.
AMBASSADOR V8 = $2775.~3095.

6 CYL. ENGS.
CONTINUE;
new 320 CID V8
(208 HP)
AVAIL. in AMBASSADOR

$2215.
STATESMAN SUPER
5545-1

$2495.
CNTRY. CLUB
5547-2
(RESTYLED)
Scena-Ramic WINDSH.

$2775.
5585-1
AMBASSADOR SUPER
new "INBOARD" HEADLIGHTS

5585-2
$2965.
57,619 BLT.

AMB. CUSTOM
208-HP V8
PACKARD ENG. OPTIONAL

(THE ONLY STATESMAN 1956)
STATESMAN SUPER
5645-1

new
Ambassador Special
(LOWER PRICED THAN AMB.)
121.3" WB

$2355. UP

130 HP
$2139. 114½" WB

14,352 (OR 17,841) BLT.

56
(6 or V8)

5665-1 (6)
↙ $2425.
AMBASSADOR SUPER 6
121.3" WB

$2939.
AMBASSADOR CUSTOM V8

5657-1

5685-2

Torque-Flo V-8

COUNTRY CLUB H/T
THE NEW
Ambassador Special
WITH new A.M.C.-BUILT V8
190 HP
250 CID
114½" WB
(INTRO. 4-56)

$2462.

(CONT'D. NEXT PAGE)

NASH

AMBASSADOR COUNTRY CLUB

PHOTOGRAPHED IN DISNEYLAND

56 $3072.
(CONT'D.)

DELUXE, SUPER, OR CUSTOM 6
REPLACE
STATESMAN 6 MODELS

AMBASSADOR SUPER
$2586.

327 CID
V·8

255 HP
$2847.

5785-1

new GRILLE and SIDE TRIM

57

THE FINAL NASH (3561 BLT.)

5787-2

AMBASSADOR CUSTOM

121½" WB

note 2-TONE SIDE TRIM

• New wider front tread for surer footing
• New sharper, easier turning
• Airliner Reclining Seats
• All-Season Air Conditioning
• Choice of Hydramatic, Overdrive or Standard
• Twin Travel Beds

5785-2

$2763.

SUPERSEDED BY
RAMBLER

JOIN THE SWING TO THE TRAVEL KING
'57 Nash
World's Finest Travel Car

new STACKED HEADLIGHTS

V8 ONLY DURING 1957, IN NASH. (6-CYL. AVAIL. IN RAMBLER.)

SEE **RAMBLER** SECTION

OLDS F-85

BY **OLDSMOBILE**
(COMPACT)

(STARTS 1961)

155 STD. HP

ENTIRE REAR DOOR RAISES, ON WAGON

61

$2621.

112" WB
6.50 × 13 TIRES
3.36 GEAR RATIO

F-85 *Cutlass*

Above: F-85 Cutlass Sports Coupe. Also available: new F-85 Club Coupe . . .

new
ROCKETTE 185 Engine
(ALUMINUM BLOCK)

185 HP
V8
(215 CID)

10.25 COMPR.
4 BBL. CARB.

$2519.
F-85 SEDAN
(4 DR.)
DE LUXE

$2713.
(WEST COAST)

$2330. ~ 2897.
PRICE RANGE

PRODUCTION:

F 85	F 85 DE LUXE
#3019 4-DR. (19,765 BLT.)	#3117 CUTLASS SPT. CPE. (9935)
3027 CLUB CPE. (2336)	3119 4 DR. (26,311)
3035 4-DR. WAGON, 6 PASS. (6677)	3135 4 DR. WAGON, 6 PASS. (526)
3045 8 PASS. (10,087)	3145 " " " 8 PASS. (757)

"...it's every inch an
OLDSMOBILE"

OLDS F-85

$2681. UP

(7909 BLT.)
$2403.

F-85 CL. COUPE

CUTLASS COUPE

(32,461 BLT.)

CUTLASS

$2694.
$2949. (WEST COAST)

(SAME PRICE AS LAST YEAR)

62

new GRILLE with "OLDSMOBILE" NAME ABOVE

2 CONVERTIBLES ADDED TO LINE : #3067 CVT. (3660 BLT.) $2760.
F85 DLX. #3167 CUTLASS CVT. (9893 BLT.) $2971.
ALSO new : F85 DELUXE JETFIRE #3147 SPT. CPE. (3765 BLT.) $3049.

$2403. ~ 3049. PRICE RANGE   215 CID V8 (155, 185 OR 215 HP)
TOTAL 1962 PRODUCTION : 102,301

F-85 COUPE
(11,276 BLT.)

TOTAL 1963 PRODUCTION : 133,522

CUTLASS

(41,343 BLT.) H/T

$2694.

$2971.

CVT. (12,149 BLT.)

$2403.

DELUXE WAGON

$3048.

(5842 BLT.)

JETFIRE H/T
(note HEAVIER SIDE TRIM)

(6647 BLT.)

$2889.

TO 195 HP with ALUMINUM V8

new SHAPE OF TAIL-LIGHT

$2592.

DELUXE SEDAN
(29,269 BLT.)

63

OLDS F-85

new GRILLE with "OLDSMOBILE" NAME ACROSS CENTER STRIP

There's 'Something Extra' about owning an OLDSMOBILE!

$2797.

WAGON (DLX.) (909 BLT.)

OLDS F-85

CUTLASS

new 230 HP

F-85 V-6 SPORTS COUPE (659 BLT.)

TOTAL 1964 PRODUCTION: 175,294

VISTA-CRS. 120" WB

WHERE THE ACTION IS!

$2938. UP

WIRE WH. COVERS AVAIL.

new VISTA-CRUISER WAGON has ROOF WINDOWS

(3394 BLT.)

64 V8 or V6

CVT.

new V6

ECON 6-WAY V-6 225 CID (155 HP)

SEDAN PROFILE

115" WB

new GRILLE

an all-new transmission

JETAWAY DRIVE

JETFIRE ROCKET V-8

TOTAL 1965 PRODUCTION: 233,154

$2937. UP

Vista-Cruiser

(5445 BLT.)

ENGINES: 225 CID V6 (155 HP) 330 CID V8 (250, 260, 315 HP) new 400 CID V8 (320 HP)

(26,441 BLT.)

CUTLASS

$2643.

65

new GRILLE; OLDSMOBILE NAME RETURNS TO CENTER

442

4-4-2 has 400 CID V8

1965

CUTLASS

The Rocket Action Car!

VISTA-CRUISER has FOLDING FORWARD-FACING 3RD SEAT

Roomy cargo area—holds over 100 cubic feet!

274

Oldsmobile CUTLASS

115" WB (EXCEPT ON VISTA CRUISER) F-85 HAS NO SIDE CHROME, PRICED FROM $2348. (6 CYL.) V8 AVAILABLE

with 12 windows.

VISTA CRUISER V8 120" WB

$2935. UP

250 cid 6 (155 HP); 330 cid V8 (250, 310 or 320 HP); 400 cid (350 HP)

SPORTS CPE.

$2633.

(13,518 BLT.)

new GRILLE WITH "OLDSMOBILE" NAME NOW ON UPPER BORDER

66

new REAR

$2348. ~ 3278. PRICE RANGE

FINAL YR. FOR 330 cid V8 (260, 310 or 320 HP)

VISTA CRUISER

$3136. UP

4-4-2 REAR

DASH

$2900.

CUTLASS SUPREME 4 DR. H/T

(22,571 BLT.)

67

new GRILLE AUX. LIGHTS BETWEEN EA. PAIR OF HEADLIGHTS

VISTA CRUISER

new 112" WB (2 DR.) 116" WB (4 DR.) 121" WB (WAGONS)

$3367. UP

68 RESTYLED

$2561. ~ 3600. PRICE RANGE

CUTLASS "S" COUPE

Drive a youngmobile from Oldsmobile

CUTLASS SUPREME 4-DR. H/T (8714 BLT.)

W-31 ENGINES: 250 cid 6 (155 HP); 350 cid V8 (250, 310 or 325 HP); 400 cid V8 (325, 350 or 360 HP)

69

GRILLES NOW SPLIT; AUX. LIGHTS NOW IN FRONT BUMPER

$3111.

A DIVISION OF GENERAL MOTORS CORP.

MODELS: F46 SPECIAL 66; G46 DYNAMIC CRUISER 76=238.1 cu 6 (100 HP)
J46 DYNAMIC CRUISER 78; L46 CUSTOM CRUISER 98=257.1 cu STRAIGHT-8 (110 HP)

1946 PRODUCTION: 114,674

SEDAN $1568. UP   76

A NEW AND FINER AUTOMATIC TRANSMISSION

GM GENERAL MOTORS   HYDRA-MATIC DRIVE

INTERIOR
$1433.
(11,721 BLT.)

66

98

46

125" WB
SEDAN (11,031 BLT.)
$1812.

CLUB SEDAN   119" WB $1407. and up, f.o.b. (66 CL.CPE.)   ($81. MORE IN '47)

(968 BLT.)   66   ENGINE SPECS. (6 and 8) AS SINCE '41   (1409 BLT.)   66

$2456.   STATION WAGON

98 CUSTOM $2307.
CRUISER CVT.

CVT.
$1681.

(3940 BLT.)

119", 125" OR 127" WB (THROUGH '48)

47

LONGER RED SECTION AROUND "OLDSMOBILE" NAME IN FRONT FENDER CHROME STRIP

It's *Smart* to own an Olds

CENTER SECTION OF BUMPER NO LONGER GROOVED AT TOP

1947 PRODUCTION: 191,454

OLDSMOBILE

$1609. *and up, f.o.b.* (66 CL. CPE. OR 2-DR.)

DELUXE SEDAN $1947.

76 (6-CYL.)

119" OR 125" WB ON OLD-STYLED 6 and 8

48

RETAINS 1947-STYLE BODY, BUT *has* "OLDSMOBILE" NAME and *new* CIRCLE EMBLEM ABOVE GRILLE, and NEW-STYLE CHROME SIDE TRIM.

new FUTURAMIC

"98" MODELS TOTALLY RESTYLED

DELUXE SEDAN

$2256.   127" WB

98 (OLDSMOBILE'S FINAL CARS *with* STRAIGHT-8 ENGINE)

"FUTURAMIC" NAME BEGINS WITH THE 8-CYL. RESTYLED 1948 OLDSMOBILES, AND IS USED FOR A FEW YEARS AFTERWARDS.

CVT. (12,914 BLT.)

2-DR. CLUB SEDAN

$2078.

$2624.

98 CONVERTIBLE *has* new HYDRAULICALLY OPERATED POWER SIDE WINDOWS *and* AUTOMATIC FRONT SEAT ADJUSTER

FUTURAMIC

MODELS :				PRICE RANGE	PRODUCTION
DYNAMIC	66	(6 CYL.)		$1609. ~ 2739.	(41,993)
"	68	8 "		1667. ~ 2797.	(16,614)
"	76	6 "		1726. ~ 1947.	(29,167)
"	78	8 "		1785. ~ 2005.	(20,651)
FUTURAMIC 98		8 "		2078. ~ 2624.	(65,335)

(EARLIEST 98s "DYNAMIC" MODELS WITH '46~'47 STYLING)

ENGINES :	IN MODELS :
6 CYL. (100 HP)	66, 76
STRAIGHT-8 (110 HP)	68, 78
" " (115 HP)	98

277

OLDSMOBILE

$1732.., and up, f.o.b.
(76 CL. CP.)

New

NEW *ROCKET* ENGINE!
(O.H.V V8)
303 C.I.D
135 HP
(TO '52)
IN 88 and 98 MODELS

76

new 105 HP
119½" wb
6

98
(125" wb)

(282,885 BLT.)

49

new AIR SCOOPS
BELOW HEADLIGHTS

new "HOLIDAY" H/T $2973.

You've got to drive it to believe it!

ALL MODELS NOW HAVE
"FUTURAMIC" BODIES :

6 - CYL. "76" = $1732. ~ 2895.
V8 "88" = $2143. ~ 3296.
V8 "98" = $2426. ~ 2973.

DELUXE STATION WAGON IS
HIGHEST-PRICED MODEL
IN 76, 88 SERIES.

"88" DESIGNATION USUALLY
ON REAR FENDER

DELUXE CLUB SEDAN
(11,820 BLT.)

$2301.

UNLIKE "98," THE new '49½ "88" has
CURVED
LOWER
EDGES
OF
WINDSHIELD.

NEW "88" (49½ INTRO.
AFTER SEASON
UNDER WAY)

LOWEST-PRICED CAR
WITH "ROCKET" ENGINE
$2375.

119½" wb

"88"

"The New Thrill"

88 DLX. SEDAN

THIS
IDENTIFIES
V8 MODELS

CVT. IS PACE CAR AT
1949 INDY 500 RACE

OLDSMOBILE ROCKETS AHEAD

FINAL 6-CYL. "76"
has NO CHROME
STRIP ON
FRONT FENDER

$1719.
and up, f.o.b.
("76" CL. CPE.)

"88"
OLDSMOBILE

88

50

new
1-PC. WINDSHIELD

88 CVT.
$2294., f.o.b.

98

8.20 x 15 TIRES
ON 98 CVT.

Make a Date
with a "Rocket 8"!

note: "98" BODY STYLE DIFFERS
FROM "76" and "88."

FINAL OLDSMOBILE
STATION WAGONS
UNTIL 1957

$1719. ~ 2772. 1950 PRICE RANGE
35 DIFF. MODELS !   PROD.: 407,889

NEW! SUPER "88"   V8s ONLY   (119½" WB ON 88 ONLY)

SUPER 88
SEDAN
(90,131
BLT.)

51

135 HP

7.60 x 15
TIRES
(88, SU-88)

$2328.

$2558.

$2049. and up (88 2-DR.)

HOLIDAY H/T
(14,180 BLT.)

(CONT'D.)

"ROCKET" **OLDSMOBILE**

SUPER 88 SEDAN
(90, 131 BLT.)
$2328.

CVT. $2673.

(34, 963 BLT.)

SUPER **"88"** 120" WB

2-DR.
$2265.

51
(CONT'D.)

"ROCKET" **98**
122" WB

New Room Inside!

DLX. HOLIDAY
H/T

$2882. (14, 012 BLT.)

$2610.
DLX.
98 SEDAN
(78,122 BLT.)

new
"HOLIDAY SEDAN"
REAR QUARTER
TREATMENT

NUMBER OF MODELS CUT TO 11

Ride the "Rocket"

1951 PRODUCTION :
285, 612

"98"

OLDSMOBILE

· 303.7 CID V8 (145 OR 160 HP)
"AUTRONIC EYE" AUTOMATIC HEADLIGHT DIMMER AVAIL.

(6402 BLT.)

88

$2262.

2-DR.

120" WB

$2462.

SU-88 SEDAN

(70,606 BLT.)

The "Rocket" Oldsmobile's New Power Steering* makes driving so easy you can...

Park with just 1 finger!

SU-88 H/T

$2673.

(15,777 BLT.)

HORIZ. GROOVES ON SU-88 FENDER PAD; VERTICAL GROOVES ON 98.

52

new VERTICAL "TOOTH" AT CENTER OF GRILLE

SUPER **88**

new SIDE TRIM (SEE DETAILS)

98 TAIL-LIGHT

Ninety-Eight

(58,550 BLT.)

SEDAN

$2786.

1952

124" WB

160 HP

new "SUPER" RANGE IN Hydra-Matic

(58,550 BLT.)

$3229.

CVT.

Ninety Eight

H/T

new SIDE TRIM DESIGN IDENTIFIES '52

98 REAR QUARTER DETAILS (SEDAN)

HOLIDAY H/T (14,150 BLT.)

$3022.

TOTAL 1952 PRODUCTION = 228,452

OLDSMOBILE

88 2-DR.

88 has 150 HP @ 3600 RPM

$2262.

(12,400 BLT.)

FINAL YR. FOR 303 CID V8s

53

HOLIDAY H/T

$2673.

(36,881 BLT.)

DETAILS OF SUPER 88 HOLIDAY H/T

DETAILS OF THE 1953 ENGINE

SUPER 88

$2462.

SEDAN (119,317 BLT.)

SU-88, 98 have 165 HP @ 3600 RPM

98 CVT. 3229.

(7521 BLT.)

Ninety-Eight

AIR COND. AVAIL.

Holiday H/T

$3022.

(27,920 BLT.)

458 98 "FIESTA" SPECIAL CONVERTIBLES ALSO BUILT, AT $5717. EACH

INTERIOR OF "98" CONVERTIBLE

POWER BRAKES and POWER STEERING ORDERED with MOST UNITS.

1953 PRODUCTION: 334,462

OLDSMOBILE

$2237., and up, f.o.b. (88 2-DR.)

new PANORAMIC WINDSHIELDS ON ALL

$2337. — **88**

$2410.

SUPER 88

170 HP @ 4000 (88)

185 HP @ 4000 (SU-88, 98)

DLX. HOLIDAY COUPÉ

$2688.

$2826. UP

ALL with new 324 CID V8s. (THROUGH '56)

54 (RESTYLED)   **98**

122" WB (88, SU-88)
126" WB (98)

Ninety-Eight

98 STARFIRE CVT.

REAR DETAILS (98)

$3248.

INTERIOR OF NEW *"Starfire"* 185 HP

LARGE, BOXY DECK AREA

1954

1954 PRODUCTION : 88 = 4-DR. (29,028 BLT.); 2-DR. (18,013); HOLIDAY H/T (25,820)
SUPER 88 = 4-DR. (111,326); 2-DR. (27,882); CONVERTIBLE (6452); DELUXE HOLIDAY H/T (42,155)
98 = 4-DR. (47,972); HOLIDAY H/T (8865); DELUXE HOLIDAY H/T (29,688); STARFIRE CVT. (6800)

$2237. ~ 3248. PRICE RANGE   TOTAL 1954 PROD. : 354,001

OLDSMOBILE

(new 4-DR. HARDTOPS JOIN EXISTING LINE OF 2-DR. H/Ts IN ALL 3 MODEL SERIES = 88, SU. 88, and 98.)

TOTAL 1955 PROD. : 583,179

$2297. ~ 3276. PRICE RANGE

"OLDSMOBILE" NAME ATOP CENTER BLADE OF GRILLE

55

HOLIDAY H/T (85,767 BLT.)

$2362., f.o.b.

SEDAN

88

$2474.

(29,028 BLT.)

7.10 x 15

88

185 HP

INT.

OLDSMOBILE'S ENTIRELY NEW

new 98 4 DR. H/T (31,267 BLT.) $3140.

SUPER 88 2-DR. (11,950 BLT.)

Holiday Sedan

A HARDTOP...WITH 4 DOORS!

7.60 x 15

$2436.

IT'S A HOLIDAY...with Sedan convenience! IT'S A SEDAN...with Holiday smartness!

Ninety Eight

NEW!

NEW!

NEW!

$3069.

7.60 x 15

DLX. HOLIDAY H/T (38,363 BLT.)

ALL-AROUND new 202 HP ENG. (SU-88, 98)

REAR.

56

FINAL YR. FOR 324 CID V8 (230 OR 240 HP @ 4400 RPM)

88

INTERIOR

88 H/T (74,739 BLT.)

$2599.

Holiday

new BISECTED GRILLE w. SMALL HORIZONTAL PCS.

7.10 x 15

$2422.

88 2-DR. (31,949 BLT.)

(CONT'D.)

284

OLDSMOBILE
SUPER 88

$2484., f.o.b.
SU-88 2-DR.

INCREASED TO $2574.

56 (CONT'D.)

7.60 x 15

98 4-DR. H/T DLX. HOLIDAY H/T SEDAN

9.48 x 15

$3456.,

(42,320 BLT.)

INCREASED TO $3551.

$2422. ~ 3740. PRICE RANGE

TOTAL 1956 PRODUCTION: 432,903 (OR 485,458)

GOLDEN ROCKET
88 CVT. (6423 BLT.)

$3182.

GOLDEN ROCKET

f.o.b.

f.o.b. PRICES START AT $2691. (88 2-DR.)

277 HP @ 4400 RPM WITH new 371 CID V8 (371 CID AVAIL. THROUGH '60)

57 (RESTYLED)

new TAIL-LIGHTS and SIDE TRIM

WAGONS RETURN!

8.50 x 14 TIRES

1957

new SUPER 88 FIESTA

$3541.

SUPER

new GRILLE

122" WB (126" ON 98)

SUPER

(8981 BLT.)

$3887.,* f.o.b.

note 3-PC. BACKLIGHT on H/T (new)

new
Starfire 98
(17,791 BLT.) H/T

4 DR. H/T (39,162 BLT.)

$3257.

*INCREASED TO $3937.

TOTAL 1957 PRODUCTION: 390,091

OLDSMOBILE

88 FIESTA WAGON (3249 BLT.)
$3284.

HOLIDAY H/T (53,036 BLT.)

88

88

$2893.

FIESTA 88

DYNAMIC 88
SUPER 88
NINETY-EIGHT
16 models to choose from!

(TOTALLY RESTYLED)

88 CVT.

122½" WB

$3262.
SUPER 88

122½" WB

$3221.↗ (4456 BLT.)

265, 305 OR 312 HP
(88) (SU 88, 98)

58

98 SEDAN (16,595 BLT.)
$3824.

(18,653 BLT.)
BADGE ON SU-88 and 98

for '58

98

126½" WB ON 98

New Rocket Engine is more powerful, gives greater performance than ever before. In addition, carburetion advances provide you with an opportunity for improved fuel savings, as much as 20%!

THE "CHROME KING" OF ALL CARS!

"OLDSmobility"

New Trans-Portable—a transistor radio that serves as your regular car radio, operating on car's built-in circuit, can also be unlocked and carried from car as a compact, lightweight portable.

New Safety-Vee Steering Wheel, with modern two-spoke, safety recessed design, allows unobstructed view of vital instrument panel gauges. New twin horn buttons are located within easy reach.

Dual-Range Power Heater gives the exact amount of heat or ventilation exactly where you want it . . . when you want it. You merely touch a button . . . power does all the work for you!

*Optional at extra cost.

DASH DETAILS →

4 HEADLTS. ABOVE new GRILLE

$2772. ~ 4300. PRICE RANGE

TOTAL 1958 PRODUCTION : 310,795

(ON ALL MODELS)
*LAVISH USE OF CHROME TRIM ON 1958 OLDSMOBILES

new "LINEAR" LOOK
LTS. SEPERATED WITHIN new GRILLE

371 CID (270 HP @ 4600)
OR
new 394 CID
(315 HP @ 4600)

59

(TOTALLY RESTYLED AGAIN)

$2837., and up, f.o.b.

(CONT'D.)

286

OLDSMOBILE

(16,123 BLT.)

$2837.

DYNAMIC 88

4-DR.

2-DR.

DY-88 HAS NO ROCKER PANEL CHROME

(LENGTH EXAGGERATED)

DYN. 88 HOLIDAY SCENICOUPE H/T (38,488 BLT.)

$2958.

59 (CONT'D.)

FIESTA (7015 BLT.)

SUPER 88 $3669.

HOLIDAY 4-DR. H/T (38,467 BLT.)

$3405.

4-DR. H/T HOLIDAY SPORTSEDANS

new 9.00 x 14 TIRES on SU-88, 98

98

(23,106 BLT.) $3890.

"CELEBRITY" ninety-eight 4-door sedan

98 CVT. (7514 BLT.)

$4366.

DASH

98 HOLIDAY SCENICOUPE H/T (13,669 BLT.)

$4086.

1959

TOTAL 1959 PRODUCTION : 366,305

$2837. ~ 4366. PRICE RANGE

OLDSMOBILE

$2900., f.o.b.
88 CELEBRITY 4-DR. SEDAN

(76,377 BLT.)

FIESTA WAGON

SUPER 88 →

H/T (16,464 BLT.)

$3325.

H/T (29,368 BLT.) → $2956.

DYNAMIC 88

4-DR. H/T (43,761 BLT.)

$3034.

60

PACE CAR AT 1960 INDY 500 RACE

new SEGMENTED GRILLE, REAR END RESTYLED

$2835. ~ 4362.
PRICE RANGE

SUPER 88

$4083.

H/T (7635 BLT.)

98

with *Roto-Matic Power Steering*

240 OR 315 HP @ 4600 RPM
(FINAL YR. FOR SMALLER (371) V8)

CVT. (5830 BLT.)

$3592.

GO OLDS '60!

DASH

SU-88 FIESTA WAGON

(7240 BLT.)

TOTAL 1960 PRODUCTION: 362,681

$3665. UP

288

power features
and accessories
for your
driving pleasure

WINDOW SWITCHES

RADIO

POWER HEATER

MANUAL HEATER

OLDSMOBILE

HOLIDAY H/T (19,878 BLT.)

$2956.

Other Oldsmobile Options include such convenience features as: Guide-Matic Power Headlight Control, Safety Sentinel, Swivel Dome and Reading Lamp, Deck Lid Power Lock Release, Electric Ventipanes, De Luxe Wheel Discs, Trim Rings and Air Conditioning.

Starglo Moroccoen interiors—optional at no extra cost in both Dynamic 88 Holiday Sedans and Holiday Coupes. And this long-wearing, easy-to-clean all-vinyl trim is as handsome as it is durable.

THIS REAR-END STYLING IN 1961 ONLY

CVT. (9049 BLT.)

$3284.

DYNAMIC **88** 2-DR. 123" WB 250 HP

(4920 BLT.)

$2835.

POWER ANTENNA

REAR QUARTER DETAIL

CLASSIC 98 HOLIDAY SEDAN

61
(TOTALLY RESTYLED)

98

DYNAMIC 88 FIESTA (AVAILABLE IN 2 and 3-SEAT MODELS)

(9387 BLT.) (ABOVE)
$3363. OR 4013.

(2 DR. 88 SEDAN)
$2835.

394 CID V8

CLASSIC 98 TOWN SEDAN

COSTLIEST '61 OLDS
$4647.

STARFIRE CVT. (7600 BLT.)

Super **88**

CVT. (2624 BLT.)
$3592.

OLDSMOBILE 123" WB

Skyrocket PERFORMANCE!

DISTINGUISHED...
DISTINCTIVE...
DECIDEDLY NEW!

new "Skyrocket"

ENGINE (394 CID V8)
325 HP @ 4600 RPM
10 TO 1 COMPRESSION
(USED IN SU-88, 98;
OPTIONAL IN DY-88)

DASH

$3325.

H/T (7009 BLT.)

"OLDSMOBILE" NAME BELOW new GRILLE (ON 88s)

Foam-padded pattern cloth, handsomely accented with lustrous Jeweltone Moroccoen, adds brilliant new sparkle to this Super 88 Holiday Sedan. Five harmonizing color choices are available.

TOTAL 1961 PRODUCTION: 253,944

#3247 HOLIDAY H/T (39,676 BLT.)

(LENGTH EXAGGERATED)

$3054. H/T

DYNAMIC 88

$3131. (53,438 BLT.)

HOLIDAY 4-DR. H/T

WAGON has UNIQUE REAR FENDER DESIGN

$2997.~ 4744. PRICE RANGE

WEST COAST, $3404., and up, f.o.b. 260 to 345 HP (THROUGH '63)

1962 PRODUCTION: 356,058

new UPRIGHT GRILLE with "OLDSMOBILE" NAME ABOVE

62 (RESTYLED)

new SIDE SCULPTURING

#3269 CELEBRITY SEDAN (68,467 BLT.)

$2997.

#3535 FIESTA WAGON (3837 BLT.)

#3569 CELEBRITY SEDAN (24,125 BLT.)

SUPER 88

$3762.

$3273.

$4744.

98 TOWN SEDAN (12,167 BLT.)

HOLIDAY SPORTS SED. (98) (33,095 BLT.)

(7149 BLT.) STARFIRE

$3984.

$4256.

98 WHEEL COVERS

98 and STARFIRE have 4 TAILLIGHTS.

(34,839 BLT.)

new HARDTOP CPE. IN STARFIRE SERIES

$4131.

290

OLDSMOBILE 63

MODIFIED GRILLES,
ALL-new TAILLIGHTS,
new SIDE TRIM

DYNAMIC 88
HOLIDAY H/T   (39,071 BLT.)
$3052.

U-88 GRILLE LIKE
DYNAMIC 88 $3408.

HOLIDAY H/T
SUPER 88   (8930 BLT.)

SU-88 FIESTA   (3878 BLT.)
$3748.

STARFIRE   REAR ROOFLINE

STARFIRE
H/T   (21,148 BLT.)
$4129.

$4742.   (4401 BLT.)
STARFIRE   REAR

98 TOWN SEDAN
(11,053 BLT.)

3982.

WHEEL COVER

98-LS
(LUXURY SEDAN)
4332.   (19,252 BLT.)

(98 DETAILS)

(23,330 BLT.)
4-DR. H/T
(SPORT SEDAN) $4238.

(INCREASED TO
$4258.)

Ninety Eight

TOTAL 1963 PRODUCTION: 371,033

291

WEST COAST
$3391.

Jetstar 88 *New full-size "88" series at a new lower price!*

new 330-cubic-inch Jetfire Rocket V-8

$2992. **'64 OLDS** *WHERE THE ACTION IS!*

(14,663 BLT.) H/T

JETSTAR 88 CELEBRITY SEDAN
$2935.

$3468. UP

FIESTA STATION WAGON (2- or 3-seat)

CELEBRITY SEDAN

SUPER 88

#3457
JETSTAR I H/T
(16,084 BLT.)

394-cubic-inch
Starfire V-8 Engine

$3069.

(32,369 BLT.)

$3062.

JETSTAR 88 HOLIDAY SEDAN
(19,325 BLT.)

JETS

$3318. Jetstar 88

DYNAMIC 88 HOLIDAY COUPE

DYNAMIC 88

CVT. (10,042 BLT.)
$3389.

J 1964

Brilliant new sports coupe in the medium-price class! **Jetstar I**

64

Starfire

#3667 CVT. (2410 BLT.)

TOWN SEDAN

(11,380 BLT.) **98**

NINETY-EIGHT

S-1964

1964

(13,753 BLT.) H/T

Ninety-Eight

OLDSMOBILE

8-64

REAR FENDER (98)

MODELS:
JETSTAR 88 = $2935.~3318.
DYNAMIC 88 = $3005.~3603.
SUPER 88 = $3256. (CELEBRITY SEDAN); $3486. (HOLIDAY 4-DR. H/T)
STARFIRE = $4138. (H/T); $4753. (CONVERTIBLE)
98 = $3993.~4468.

ENGINES: 330 CID V8 (230, 245 OR 290 HP)
394 " " (280, 330 OR 345 HP)

TOTAL 1964 PRODUCTION: 335,637

OLDSMOBILE

$3072.
JETSTAR 88
HOLIDAY 4 DR. H/T
(15,922 BLT.)

330 CID V8   260 HP   123" WB
7.75 × 14 TIRES
Jetstar 88

DYNAMIC 88 LINE
JOINED BY
new DELTA 88 →

← *new* STYLING!

(22,725 BLT.) $2938.

Delta 88.

(23,194 BLT.) ↗
$3253.

8.25 × 14
TIRES

(24,746 BLT.) $3065.

DYNAMIC
↙ 88
H/T

123"
WB

425 CID
SUPER
ROCKET
V8
360 HP
(TO 370 IN DELTA
88)

GC·5580

new GRILLE

DELTA 88 DASH

$2938. ~ 4778.
PRICE RANGE

330 CID V8
(260 HP)

425 CID V8
(310, 360
OR 370 HP)

65

TOTAL 1965
PRODUCTION:
400,664

note THAT STARFIRE *and* 98
have OWN GRILLE
DESIGNS

370 HP
STARFIRE → (13,024 BLT.)
H/T

(2238 BLT.)   CVT.

$4761.

98 (126" WB)

$4334.,
f.o.b.

98
LUXURY
SEDAN

8.55 × 14 TIRES

WITH
VINYL
TOP

GS 4201

$4138.   WEST
COAST $4237.

Ninety-Eight

$4273.
HOLIDAY
SPORT
SEDAN
(28,480 BLT.)

LS

NINETY-EIGHT

IIM 4380

98 DASH

'65 OLDSMOBILE
The Rocket Action Car!

CUTLASS, F-85 and TORONADO MODELS ILLUSTRATED SEPARATELY.

Oldsmobile

$2927.~4443. PRICE RANGE

(WEST COAST PRICES IN SMALL PRINT)

GM

GENERAL MOTORS

$3328.

4-DR. H/T HOLIDAY SEDAN $3757.

JETSTAR 88 FROM $3314.
DYNAMIC 88 FROM $3442.

123" WB

88

(33,326 BLT.)

DELTA 88

$3253.

2-DR. H/T HOLIDAY CPE. $3682.

(20,857 BLT.)

66

123" WB
425 CID V8
(365 HP)

Starfire:

425 CID V8 (310 HP)
STD.
(JETSTAR 88 has
330 CID V8
(260 HP)
(13,019 BLT.)

$3564.

(STARFIRE IS LUXURY H/T WITH BIG V8. NOT TO BE CONFUSED WITH SMALL CAR OF 1975 ON, WITH SAME NAME.)

Ninety-Eight:

98 SEDANS FROM $4592.

126" WB
(THROUGH '68)

98 has CRISS-CROSS GRILLE PCS.

$3966.

STEP OUT FRONT IN '66... in a Rocket Action Olds!

365 HP (98)

ENGINES : 330 CID V8 (250 or 260 HP) ; (JETSTAR 88)
330 CID V8 (320 HP) (OPT. IN JETSTAR 88)
425 CID V8 (300, 310, 365 or 375 HP) (DYNAMIC 88, DELTA 88,
STARFIRE or 98)

1966 PRODUCTION :
318,667 (88s, 98s)
(578,385, INCL. F-85,
CUTLASS, TORONADO)

Oldsmobile

(new DELMONT 88 REPLACES FORMER JETSTAR 88)

TOWN SED. $3543. (WEST)

330 OR 425 CID V8s (250 TO 375 HP)

$3008. UP

4-DR. H/T $3675. (WEST)

has DELMONT 88 NAME OVER COWL

IN "330" OR "425" SERIES:

DELMONT 88
Brand-new 88 series!
Goes to show what Olds can do
with a modest price tag . . .
and a lot of Toronado inspiration.

DELMONT 88 DASH

8.55 x 14 TIRES

CVT. (3525 BLT.) (ONLY AVAIL. IN "425" SERIES) $3462.

$3912. (WEST)

1967 PRODUCTION: 277, 910 (88s and 98s)

67 new GRILLES

$3008. ~ 4498. PRICE RANGE

Engineered for excitement . . . Toronado-style!
'67 OLDSMOBILE

The Rocket Action Cars are out front again!

(21,909 BLT.) (STD.)

3386.

(INTRO. 9-29-66)

$3646.

(2447 BLT.)

DELTA 88 CVT.

DELTA 88 4-DR. H/T HOLIDAY SEDAN $3954. (WEST)

$3218. (WEST)

DELTA 88 $3786.

REAR

(CUSTOM 2 and 4 DR. H/Ts ALSO)

(22,770 BLT.)

TOWN SEDAN

DELTA 88 HOLIDAY COUPE (2-DR. H/T) $3878. (WEST)

$3310 STD.

(CONT'D. NEXT PAGE)

STD. (14,471 BLT.)

Oldsmobile

DELTA 88 CUSTOM MODELS DIFFER THUS FROM STANDARD DELTA 88s: CUSTOM MODELS' SIDE TRIM

How do you top a line of cars as luxurious as the Olds Delta 88 for 1967?

Bring on an all-new, ultra-luxurious Delta 88 Custom.

HOLIDAY H/T $3522.

(12,192 BLT.)

CLOSE VIEW OF SIDE TRIM LOUVRES

(14,306 BLT.)

425 CID V8

DELTA 88 CUSTOM
Two all-new Custom hardtops highlight the Delta 88 line.

1967

CUSTOM 4 DR. H/T

$4011. (WEST)

$3582.

67 (CONT'D.)

Ninety-Eight:

HOLIDAY CPE. (H/T)

$4736. (WEST)

$4214. (10,476 BLT.)

98 LUXURY SEDAN (35,511 BLT.)

$4351.

$4873. (WEST)

$4498.

98

365 HP

note VERTICAL TAIL LTS.

CLOSE-UP (REAR)

(98)

$5020. (WEST)

CVT. (3769 BLT.)

8.85 x 14 TIRES

98 DASH

$4798. (WEST)

HOLIDAY H/T SEDAN

(17,533 BLT.)

$4276.

Oldsmobile

(INTRO. 9-21-67)

FINAL YEAR FOR **DELMONT 88** ➚

88 DELTA GRILLE NAME)

H/T (18,391 BLT.)

$3202. ➚

Tilt & Telescope steering wheel"

$3146. ~ 4618. PRICE RANGE

ENGINES : 330 CID V8 (250, 260 or 320 HP) 425 CID (300, 310, 365 or 375 HP)

Drive a youngmobile from Oldsmobile

68

new SPLIT GRILLES

$4497.

V8s UP TO 455 CID

98

LUXURY SEDAN (40,755 BLT.)

1968 PRODUCTION : 331,566 (88s and 98s) 627,533 (ALL MODELS)

Escape from the ordinary in Olds

DELTA 88 has 350 CID V8 (250 HP)

69

new GRILLES

Delta 88 Royale (new 124" WB) ➝ (INTRO. 9-26-68)

(310 HP) 455 CID V8

REAR

8.55 x 15 TIRES

Delta 88 Royale

note SIDE LOUVRES

OLDSMOBILE NOW SHOWING YOUNGMOBILE THINKING 1969

98 REAR

Olds Ninety-Eight (new 127" WB) ➝

HOLIDAY H/T

27,041 BLT.) $4461.

$3222. ~ 4719. PRICE RANGE

(BRIGHTWORK ON ROCKER PANELS)

1969 PROD. : 368,045 (88s and 98s) ; 635,241 (ALL MODELS)

Oldsmobile TORONADO (SINCE 1966)

Front-wheel drive

$4617. UP

$5125. UP (WEST COAST)

NEW 66

119" WB (THROUGH '70)

(40,963 BLT.)

STEP OUT FRONT IN '66,...in a Rocket Action Olds!

TORONADO

V8 ENGINE (ON ALL)
425 CID
385 HP

(21,790 BLT.)

new GRILLE   new WHEEL COVERS

67

(STD. OR DLX. 2-DR HT, 1966 and 1967)

$5182. UP (WEST COAST)

$4674. UP

$4750. $4945. CUSTOM

$5258. (WEST COAST)

(26,454 BLT.)

Toronado. Test drive the front-wheel-drive "youngmobile" from Oldsmobile.

68

new SPLIT GRILLE

new 455 CID V8 (375 HP

OLDSMOBILE NOW SHOWING YOUNGMOBILE THINKING 1969

(28,494 BLT.)

Escape from the ordinary

69

new REAR DECK

MODIFIED GRILLE

AS BEFORE, 8.85 x 15 TIRES

$4835. $5030. (CUSTOM)

$5344. (WEST COAST)

PACKARD MOTOR CAR CO., DETROIT
(STUDEBAKER-PACKARD CORP.,
1954~1958)

PACKARD

(1899 TO 1958)

new CLIPPER 8 CYL.

(STARTS 4-41)

41½

(1951 SERIES)

$1375.

OTHER 1941 MODELS CONTINUE ALSO

Clipper

new 2-DR. CLIPPERS NOW ALSO AVAIL.

110

180

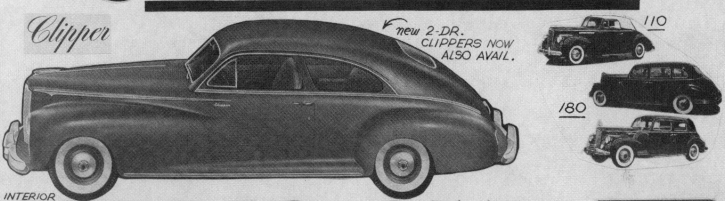

INTERIOR

LOOKING AHEAD? SKIPPER THE CLIPPER

42-45

(2000 SERIES)

6776 BLT. '42. 2652 AVAIL. 1943 TO 1945 FOR MILITARY STAFF

ELECTROMATIC DRIVE

SIMPLIFIED DRIVING WITH NO JERK·NO SLIP·NO CREEP

new CHOICE of 6 OR 8-CYL. CLIPPERS

SUPER 8 *has* 148" WB

CUSTOM SUPER CLIPPER

$1746.
DELUXE CLIPPER

46-47

(2100 SERIES)

(BIG-CITY PACKARD SHOWROOMS MORE LUXURIOUS THAN THIS RURAL OUTLET)

42,102 BLT. 1946

55,477 BLT. 1947

PACKARD

$3161.   EIGHT

STATION SEDAN
(new)

NEW   SMOOTH
SIDE BODIES

SUPER-8 CVT.
IS FIRST OF
1948 PACKARDS
TO BE INTRODUCED.

$2990.

SUPER 8
130 HP

EARLY

48-49
(RESTYLED) (2200 SERIES)

$2529.

$3461.   CUSTOM 8
160 HP

127" WB

$3866.

1948 - EARLY '49 DASH
ILLUSTR. ON NEXT PAGE

CUSTOM 8 *has*
CRISS-CROSS PIECES
IN GRILLE.

ASK THE MAN
WHO OWNS ONE

3 DIFF. STRAIGHT-8 ENGINES : 288 CID (130 HP) = 8, DELUXE 8
327 CID (145 HP) = SUPER 8   1948 = 98,897 BLT.
356 CID (160 HP) = CUSTOM 8   1949 = 104,593 "

PACKARD

CLOSE-UP VIEW OF DASH (2200 SERIES)

135 HP (8)
150 HP (SU. 8)
160 HP (CUST. 8)

$2383. ('50)

DELUXE 8

$2224. UP (LOWEST-PRICED MODEL)

CLUB SEDAN

EIGHT (120" WB)

$2633.

DASH and BACKLIGHT DETAILS

SUPER 8 new 127" WB

Ultramatic Drive AVAIL.

"Golden Anniversary" MODELS
new LARGER BACKLIGHTS ON 4-DOOR SEDANS

49½-50 (2300 SERIES)
77 MAJOR IMPROVEMENTS

CUSTOM 8 127" WB

STATION SEDAN (WAGON) STILL AVAIL. (THROUGH '50)

#2333 CONVERTIBLE (68 BLT. '49½; 77 BLT. '50)
$4295. (1949½) $4520. (1950)
(HIGHEST-PRICED MODEL)

FINAL 356 CID STR. 8 (CUSTOM 8)
(288 and 327 CID 8s CONTINUED THROUGH 1954.)

new: SIDE TRIM ADDED:
HORIZONTAL CHROME STRIP and "PACKARD" NAME

TOTAL 1950 PRODUCTION: 72,138

301

PACKARD

200 CLUB SEDAN **$2366.** UP

250 (CVT. and H/T) (4640 BLT. '51; 5201 BLT. '52)

Prestige car of the medium-priced field: Packard '200' Club Sedan—$2366"

—one of nine exciting new models for '51

122" WB

250 CVT.

"*NEW, ALL-NEW*"

CVT. ('51)
$3391.
CVT. ('52)
$3476.

$3234. UP

250 MAYFAIR

51-52

(TOTALLY RESTYLED 2400 SERIES)

(2500 SER.)

300

$3034. UP

REAR SIDE DETAILS

(15,309 BLT. '51; 6705 BLT. '52)

1952 MODEL ← (left) SIMILAR, BUT has new HOOD ORNAMENT and MEDALLION ON GRILLE

Ultramatic

400 PATRICIAN

COSTLY MODELS CONTINUE CORMORANT FIGURE AS 1951 ORNAMENT →

(9001 BLT. '51)
(3975 BLT. '52)

1951 MODELS have "PACKARD" NAME ABOVE GRILLE

PACKARD

New Armor-rib body construction!
New Tele-glance instrument panel!
New Safeti-set brake!

—the one for '51!

$2302. ~ 3662. (1951 PRICE RANGE)
$2494. ~ 3797. (1952)
(200) (400 PATRICIAN)

ENGINES: (STRAIGHT-8)
288 CID (135 HP)
327 CID (150 or 155 HP)

76,075 BLT. 1951

62,988 BLT. 1952

PACKARD

(STANDARD) CLIPPER PRICED FROM **$2544.** (2-DR. CLUB SEDAN) (6370 BLT.) #2695

$2588. SEDAN (26,027 BLT.) (INCR. TO $2745.)

CLIPPER DELUXE   122" WB   #2662

#2665 CLUB SEDAN (4678 BLT.) **$2691.**

new HOOD ORNAMENT and SMOOTH HORIZONTAL GRILLE PIECE (ON CLIPPER ONLY)

New Packard **CLIPPER**   160 HP

53 (2600 SERIES)

CLIPPER SERIES RETURNS (PREVIOUSLY AVAIL. 1941 – 1947)

new CAVALIER   127" WB   (10,799 BLT.)   #2672   **$3234.**

#2679   (..T. ..8 BLT.)   **$3486.**

MAYFAIR   122" WB

MAYFAIR H/T ALSO, (5150 BLT.), W/O 3 CHROME REAR FENDER PLAQUES SEEN ON ABOVE CVT.

(25 CUSTOM BLT. #2653 DERHAM SEDANS AT $6539.)

$3740.

(150 LIMOS., 8 PASS. EXECUTIVE SEDANS, ON 149" WB)

400 PATRICIAN SEDAN #2652 (7456 BLT.) 127" WB

new GROOVES IN HORIZONTAL GRILLE PIECE (EXCEPT ON CLIPPER)

..ote "CONTINENTAL" REAR ..PARE TIRE and ..OVER

new CARIBBEAN   #2678

$5209.

(750 BLT.)

STRAIGHT-8 ENGINES: 288 CID (150 HP) (CLIPPER)
327 CID (160 HP) (" DELUXE)
327 CID (180 HP) (CAVALIER, MAYFAIR, CARIBBEAN)
(9-MAIN-BEARING 327 CID ENG. IN PATRICIAN, LIMOUSINES)

TOTAL 1953 PRODUCTION: 80,371

303

PACKARD

RARE CLIPPER SPECIAL (970 SEDANS, 912 CLUB SEDANS BLT.)
$2594. $2544.

$3125. (3618 BLT

#5467 → PANAMA

CLIPPER DELUXE $2645. UP

SUPER CLIPPER

122", 127" OR 149" WB
150, 165, 185 OR 212 HP

54

(5400 SERIES)

SEDAN (6270 BLT.) $2815.

$3344.

DASH

CAVALIER 127" WB
#5472 SEDAN (2580 BLT.)

PATRICIAN 127" WB

TOTAL 1954 PRODUCTION: 27,307
FINAL STRAIGHT-8 ENGINES
(288, 327 OR 359 CID)

$3740.

STUDEBAKER-PACKARD MERGER

(7456 BLT.)

122" WB

CLIPPER WITH (FINE VERTICAL GRILLE PCS.)

FIRST MAJOR RESTYLING SINCE 1951

PATRICIAN

PRICE RANGE: $2586. TO $5932.

TOTA 1955 PRODUCTION

68,674 EST.

127" WB $3740.

new 12-VOLT ELECTRICAL SYSTEM
CARIBBEAN CONVT. (APPROX. 490 BLT.)

400

new V8 O.H.V. ENGINES! (4600 RPM)
320 CID (225 HP) (CLIPPER DLX., SUPER)
352 CID (245 HP, CLIPPER CUSTOM)
(260 HP, PACKARDS)
(275 HP, CARIB.)

5500 SERIES

55

TOTALLY RESTYLED

$5932.

PACKARD

(CLIPPER TREATED AS A SEPARATE MAKE FROM PACKARD IN 1956.)

CLIPPER SUPER also avail. #5642, 5647

2731. #5622 SEDAN

CLIPPER DELUXE (5715 BLT.)

$3069. ← #5662 SEDAN (2129 BLT.)

(SUPER SEDAN, H/T ALSO) $2866., $2916.

MEMBERS of CLIPPER GRILLE NOW HORIZONTAL.

CUSTOM CLIPPER

CUSTOM CLIPPER CONSTELLATION H/T → #5667

$3164. (1466 BLT.)

122" WB (CLIPPERS) OTHERS, 127" WB

56 new GRILLES

(5600 SERIES)

New DISPLACEMENT of 374 CID on ALL PACKARD V8 ENGINES. ALL BUT CARIBBEAN have 290 HP @ 4600 RPM.

PACKARD MODELS

$3483. (1784 BLT.)

122" WB

EXECUTIVE

H/T (1031 BLT.)

$3658.

WIDER-SPACED GRILLE PIECES with MESH BACKGROUND

$4160.

127" WB PATRICIAN (3775 BLT.)

$4190. 400 H/T

(3224 BLT.)

CARIBBEAN has 310 HP @ 4600 RPM

"ASK THE MAN WHO OWNS the New ONE"

$5995.

(276 BLT.)

263 CARIBBEAN H/Ts ALSO, $5495. (A SMALL NO. OF '55 H/Ts DISCOVERED!)

1956 PRICE RANGES =
$2731. ~ 3164. (CLIPPER)
3465. ~ 5995. (PACKARD)

TOTAL 1956 PRODUCTION: 13,432, PLUS 18,482 CLIPPERS.

PACKARD
(57-L SERIES)

57-L
57
Y8 SEDAN

new 120½" WB

$3212.

CLIPPER

SEDAN and WAGON are ONLY CHOICES LISTED DURING 1957.

new SMALLER V8 DISPLACEMENT OF 289 CID

HP REDUCED TO 275 @ 4800 RPM

275 HP (THROUGH '58)

(3940 BLT.)

new 116½" BODIES LIKE STUDEBAKER (THROUGH '58)

P8

#P8 WAGON (159 BLT.)

$3384.
(AS IN '57)

(58-L SERIES)
58
THE FINAL PACKARDS

58-L

$3212.

See the all-new '58 Packards:
- The panoramic Packard Hardtop
- The luxurious Packard 4-door Sedan
- The supercharged Packard Hawk
- The versatile Packard Station Wagon

Studebaker-Packard
CORPORATION
Where pride of Workmanship comes first!

#J8 (1200 BLT.) SEDAN

ENG. SPECS. AS IN 1957. 210 HP, BUT 275 HP IN new #K9 HAWK H/T (ILLUSTRATED)

4 HEADLIGHTS (EXCEPT ON HAWK)

FRONT END DETAILS

$3995. (588 BLT.)

HAWK has 2 HEADLIGHTS, and A LOWER GRILLE

non-HAWK H/T ALSO (675 BLT.)
$3262.

(PACKARD DISCONTINUED 1958)

TOTAL PRODUCTION, (1957 = 5495)
1958 = 1745
(ONLY 4 MODELS IN 1958)

48 HP with 133 CID HERCULES ENGINE OR 40 HP with 91 CID CONT. ENG.

STEEL RETRACTABLE TOP

97 BLT. 4 CYL.

PLAYBOY
PLAYBOY MOTOR CAR CORP., BUFFALO, N.Y. (1946-1951)

48

3.73 OR 4.1 GEAR RATIO 90" W.B.

$985.

INTER.

1946 MODELS HAVE 16" WHEELS PAINTED BODY COLOR WITH CONTRASTING STRIPING.

3-YEAR BODY TYPE PROD. FIGS.:
(49,918 DLX.; 125,704 SPL.DLX.)

2 DR. SEDAN

Plymouth

(A DIVISION OF CHRYSLER CORP.)
(SINCE MID~1928)

STATION WAGON (WOODEN BODY)

SPECIAL DE LUXE
has
CHROME EFFECT
ON WINDSHIELD
FRAME

(12,913)

CVT. (15,295)

1946 has FLAT BUTTON TYPE DOOR LOCK COVERS.

4 DR. SEDAN
(120,757 DLX.;
514,986
SPL.
DLX.)

1948 has new
7.50 × 15
LOW-PRESSURE TIRES.

46-48 *

P-15S DLX. or P-15C SPECIAL DLX.

95 HP @ 3600 RPM

WHITE BEAUTY RINGS ADDED 1947, DURING MODEL YR.

* = CONT'D. TO 2-49

(10,400 DLX.;
156,629
SPL.DLX.)

CLUB COUPE

$1075. TO $2068.
(PRICE RANGE, 1946 TO EARLY 1949)

CONVERTIBLE DASH IS PAINTED IN BODY COLOR, INSTEAD OF being WOODGRAINED.

REAR (SEDAN)

SEDAN INTERIOR

SPECIAL DE LUXE has RADIO GRILLE

DE LUXE

(3-PASS. BUSINESS COUPE ALSO AVAIL., with SMALLER REAR QUARTER WINDOWS and SHORTER CAB THAN CLUB COUPE.)
(16,117 DLX.; 31,399 SPL. DLX.)

P-15 PLYMOUTH PRODUCTION : 1945 (YEAR'S END) = 749
1946 = 241,656
1947 = 347,946
1948 = 381,139

Plymouth

$1629.

SPEC. DLX. SEDAN + INTERIOR

new ALL-METAL 2-DR. SUBURBAN WAGON $1840.

P-17

SEDAN REAR DOORS NOW FRONT-HINGED

6.40 × 15 or 6.70 × 15 TIRES (THROUGH '52)

SPECIAL DE LUXE 4-DOOR WAGON has WOODEN PANELS.

P-18 $2372.

$1982.

new 97 HP @ 3600 RPM (THROUGH '52)

SPEC. DLX. CLUB COUPE

$1603.

new "Double-Size" CVT. BACKLIGHT has REMOVABLE, ZIPPERED CENTER SECTION

3-WINDOW BUSINESS CPE. AVAIL., $1371.

49

P-17 (111" WB)
P-18 (118½" WB)

(TOTALLY RESTYLED)

HORIZONTAL CREASES on BUMPERS ('49 ONLY)

new SWITCH-KEY STARTING

SLOGAN : "The car that likes to be compared"

PRODUCTION : **P17 DELUXE** = BUSINESS CPE. (15,715); 2-DR. (28,516); 2 DR. SUBURBAN WAGON (19,220);
P18 DELUXE = CLUB CPE. (25,687); 4-DR. SEDAN (61,021)
P18 SPECIAL DELUXE = CLUB CPE. (99,361); 4-DR. SEDAN (234,084); CONVERTIBLE (12,697);
4 DR. WAGON (2059)

PRODUCTION = MODEL YEAR 498,360 ; CALENDAR YEAR 569,260

Plymouth

2-DR. (67,584 BLT.) $1371.

DE LUXE

3-WINDOW BUSINESS COUPE

(16,861 BLT.)

$1492.

SUBURBAN

CLUB CPE. (99,361 BLT.)

4 DR. WAGON

SPECIAL DE LUXE

$1603.

$1629.
(234,084)

PRICE RANGE: $1371. TO $2372.

$1982.

(12,697 BLT.)

new EMBLEM

P-19 DE LUXE (111" WB)

P-20 DE LUXE; SPEC. DLX. (118½" WB)

50

PLYMOUTH

DASH

new SMOOTH BUMPER SURFACE

new GRILLE has FEWER PIECES.

MODEL LINE-UP CONTINUES AS IN 1949, BUT SUBURBAN 2-DR. ($1840.) IS JOINED BY new SUBURBAN 2-DR. SPECIAL ($1946.) TOTAL NUMBER PRODUCED OF BOTH IS 34,457. ALSO, 2059 4-DR. WOOD-PANELED WAGONS BUILT (AT $2372.)

TOTAL 1950 PRODUCTION = 567,381

Plymouth (76,250)

P-22 CONCORD

(49,139) 111" WB 2-DR.

$1537.~2222.
1951 PRICE RANGE
(14,255) 3-WINDOW COUPE

(388,785)

(15,650)

REAR DETAILS 1951

1951 MODELS ILLUSTRATED UNLESS OTHERWISE NOTED.

1951 BELVEDERE is new H/T.

(51,266)

SHIELD BADGE REPLACED BY CIRCLE ON '52.

MODEL NAME in SCRIPT ON 1952 FRONT FENDER

P-23 CAMBRIDGE and CRANBRK. have 118½" WB
$1610.~2329.
1952 PRICE RANGE

51-52

new CONCORD, CAMBRIDGE, CRANBROOK MODEL NAMES

('51)

DASH

'50 PLYMOUTH TAXI

1952 BELVEDERE (BELOW) has new REAR COLOR SWEEP

('52)

TOTAL PLYMOUTH PRODUCTION, 1951 = 607,691
"   "   " 1952 = 466,289

Plymouth

new 100 HP @ 3600 RPM

SAVOY WAGON

new SPORT WIRE WHEELS OPTIONAL

CRANBROOK BELVEDERE

53

(TOTALLY RESTYLED)
P-24-1 CAMBRIDGE
P-24-2 CRANBR.

CRANBROOK

CRANBROOK

new 114" WB (THROUGH '54)

6.70 x 15 TIRES (TO '56)

$1618.~2220. PRICE RANGE

TOTAL 1953 PRODUCTION = 654,414

P-25-3 BELVEDERE

P-25-1 PLAZA

DASH

LATE '54 has *new* 230.2 CID and 110 HP @ 3600 RPM

P-25-2 SAVOY

$1618. UP

54

EARLY 1954 BELVEDERE H/T DOES NOT HAVE THIS COLOR BAND ON SIDE

BELVEDERE

(25,592 BLT.)

$1618.~2301. PRICE RANGE

TOTAL 1954 PRODUCTION = 396,702

Plymouth

230 CID 6 CYL. OR new 241 CID or 260 CID V8s.

$2077. UP
2-DR. WAGON

new AUTOMATIC TRANSMISSION CONTROL on DASH

2-DR. and 4-DR. WAGONS in PLAZA SERIES.

PLAZA

$1738. UP

SAVOY
(NO SAVOY WAGONS) $1880. UP

55

6 = 117 HP @ 4000 RPM

V8 = 157 OR 167 HP @ 4400 RPM

(TOTALLY RESTYLED with new "FORWARD LOOK")

new 115" WB (THROUGH '56)

CLUB COUPE

new PANORAMIC WINDSHIELD

$1738. ~ $2425. PRICE RANGE

$2322. UP

6 CYL. = $1936.
V8 = $2039.

(41,645 BLT.)

BELVEDERE
6 OR V8 IN ALL MODEL SERIES

4 DR. WAGON ONLY IN BELVEDERE SERIES

BELVEDERE SUBURBAN WAGON (18,488 BLT.)

BELVEDERE H/T (47,375 BLT.)

1955 PRODUCTION: 746,361

$2113. UP

6-CYL. HAS STRAIGHT EMBLEM ABOVE GRILLE →

REAR VIEW

1955

new FRENCHED HEADLIGHTS

V8 HAS ABOVE TYPE OF EMBLEM

312

Plymouth

$2196. (23,866 BLT.) $2314. (33,333 BLT.)

6.70 × 15 TIRES (ALL BUT new FURY)

SUBURBAN
DE LUXE 2-DR. WAGON

$2267. CUSTOM SUBURBAN 4 DR. / 2 DR.

PLAZA

(9489 BLT.)

$1784. UP

$2109.
84,218 BLT.

BELVEDERE
SEDAN (ABOVE)

$2484. SPORT SUBURBAN (15,104 BLT.)

SAVOY   H/T

$2130.   (16,473 BLT.)

56 P-28 (6)
      P-29 (V8)

PUSHBUTTON POWERFLITE:

new BELVEDERE 4-DR. H/T (BELOW)

7,515 BLT.

ENGINES:
230.2 CID 6 (125 OR 131 HP)
270 CID V8 (180 HP) (1956 ONLY)
277 CID V8 (187 OR 200 HP)
(new FURY has OWN 303 CID V8)

TOTAL 1956 PRODUCTION:
552,577

2287.

new **Fury**

240 OR 270 HP (WITH new 303 CID V8) 7.10 × 15 TIRES)

$2866.   (4485 BLT.)

new SHARPLY-PEAKED TAIL FINS

H/T

new MESH AT GRILLE CENTER

1956

Plymouth

PLAZA — $2009.

$2229. H/T (31,373 BLT.) — SAVOY

BELVEDERE — $2349. ↑ H/T

318 CID V8 IN FURY →

new 8.00 × 14 TIRES ON FURY H/T $2925.

EARLY '57 (6 OPEN SLOTS BELOW BUMPERS) ←

2 DR. (49,137 BLT.)

4 DR. H/T $2419. (67,268 BLT.)

(7438 BLT.)

new 118" WB (122" WB ON WAGONS) (THROUGH '61)

SPORT SDN. (BELV.)

BELVEDERE

LATE '57 (EXTRA VERTICAL MEMBERS BELOW BUMPER) ←

TAILGATE WINDOW DETAIL

SECRET LUGGAGE COMPARTMENT. Almost 10 cubic feet of locked space for safe, out-of-sight storage of luggage, cameras and other valuables. On all 6-pass. models.

$2622. UP (23,402 BLT.)

new 7.50 × 14 TIRES (ALL BUT FURY)

57

(TOTALLY RESTYLED)

P-30 (6)
P-31 (V8)

EARLY '57 FRONT END CLOSE-UP

DASH

$1899. ~ 2777. PRICE RANGE

TOTAL 1957 PRODUCTION : 655,006

SLOGAN : "SUDDENLY IT'S 1960!"

ENGINES : 230.2 CID 6 (132 HP)
277 CID V8 (197 OR 235 HP)
301 CID V8 (215 OR 235 HP)
318 CID V8 (290 HP) (FURY)

Plymouth

SUBURBAN

$2432.

(15,625 BLT.) The De Luxe Suburban—2-door, 6-passenger

$2553.

(5925 BLT.) The Custom Suburban—2-door, 6-passenger

The Custom Suburban—4-door, 9- or 6-passenger

SPORT SUBURBAN

(23,170 BLT.) $2760. UP

← Star of the Forward Look

PLAZA $2028.

(1958 IS FINAL YEAR FOR PLAZA SERIES.)

The Plaza 2-door Business Coupe

(1472 BLT.)

SAVOY (67,923 BLT.)

$2305.

The Savoy 4-door Sedan

$2400. The Savoy 4-door Hardtop

(5060 BLT.)

(49,124 BLT.)

$2440. **BELVEDERE**

The Belvedere 4-door Sedan

$2762. →

The Belvedere Convertible

(9941 BLT.)

58 NEW GRILLE

LP-1 (6)
LP-2 (V8)

$2028.~2900.
PRICE RANGE

TOTAL 1958 PRODUCTION:
366,758

INSTRUMENT CLUSTER

$2457.

(36,043 BLT.)

7.50 × 14 TIRES

BELVEDERE

(8.00 × 14 ON 9-PASS. WAGONS AND FURY H/T)

← FURY

(5303 BLT.)

$3067.

230 CID 6 (132 HP @ 3600)
318 CID V8
(225 OR 250 HP @ 4400)
350 CID V8
(305 OR 315 HP
@ 5000 RPM)

4 HEADLIGHTS

newest engine—"Golden Commando V-8"

(WITH ELECTRONIC FUEL INJECTION)

SILVER SPECIAL (RARE!)
(PLAZA)

NOTE SPECIAL SIDE TRIM

315

Plymouth

CUSTOM SUBURBAN

SAVOY
4-door Sedan, V-8 or 6

BELVEDERE
2-door Sedan, V-8 or 6

$2881. UP

OPTIONAL
new
SWIVEL
SEATS
(STD. IN SPORT FURY)

59
MP-1 (6)
MP-2 (V8)

7.50 x 14 TIRES

DASH

SPORT SUBURBAN

(2-DR. and 4-DR. WAGONS IN DELUXE and CUSTOM SUBURBAN SERIES.)

6 PASSENGER (7224 BLT.) = $3021.
9 " (9549 ") = $3131.

(NO MORE LOW-PRICED PLAZA SERIES.)

$2143. ~ 3131.
PRICE RANGE

TOTAL 1959 PRODUCTION: 393,213

new DECORATIVE "SPARE TIRE COVER" ON DECK LID OF SPORT FURY

$2927.

new GRILLE

new SPORT FURY H/T
(17,867 BLT.)

FINAL USE OF L-HEAD DESIGN IN PLYMOUTH SIX

new TAIL FINS

CVT.
(5990 BLT.)

SPORT FURY CONVERTIBLE

ENGINES:
230 CID 6 (132 HP @ 3600)
318 CID V8 (230 or 260 HP @ 4400 RPM)
361 CID V8 (305 HP @ 4600 RPM)

SEDAN $2439.

Plymouth

SAVOY

BELVEDERE

H/T $2641.

V8s have 318, 361, OR 383 CID (230, 260, 305, 310, 325 OR 330 HP)

new SLANTING O.H.V. 225 CID 6 (145 HP @ 4000 RPM) (TO '71)

note THE REAR FENDER ORNAMENTS WHICH IDENTIFY EACH INDIVIDUAL MODEL SERIES.

(9036 BLT.) $2656.

FURY

CUSTOM SUBURBAN $2880. UP

4-DR. H/T

2-DR. H/T

WITH SEMI-RECTANGULAR STEERING WHEEL

WITHOUT GRILLE GUARD

WITH GRILLE GUARD

new GRILLE

60

PP-1 (6 CYL.)
PP-2 (V8)

SHOWN with ROUND STEERING WHEEL

DASH

VALIANT COMPACT SERIES INTRO.

CLOSER DETAILS OF WAGON

7.50 x 14" TIRES

TOTAL 1960 PRODUCTION: 447,724 (252,453 FULL-SIZED PLYMOUTHS)

SEE ALSO: VALIANT

$2260. ~ 3134. PRICE RANGE

FROM 1960 ON, VALIANT PRODUCTION FIGURES LISTED SEPARATELY.

Plymouth

7.00 × 14 TIRES (6)
7.50 × 14 ON 6-CYL. WAGONS and V8s.
8.00 × 14 ON 9-PASS. V8 WAGON

ALTERNATOR TEST
DETROIT to CHICAGO

9-PASS. V8 IS COSTLIEST MODEL, AT $3134.

PLYMOUTH—This car traveled 328 miles without a battery. Alternator, standard on 1961 Chrysler Corporation cars, provided all necessary electrical energy.

61
RP-1 (6)
RP-2 (V8)

SPORT SUBURBAN

SUBURBAN
2 and 4 DR. WAGONS
$2602. UP

FURY 4-DR. H/T $2656. (6)
(8507 BLT.) $2775. (V8)

TOTAL 1961 PRODUCTION :
188,170 (FULL-SIZED)

ENGINES :
225 CID SLANT-6 (145 HP)
318 CID V8 (230 or 260 HP)
361 CID V8 (305 HP)
383 CID V8 (330 HP)
ALSO,
LARGEST OF 4 PLYMOUTH V8s
IS NEW 413 CID ENGINE (UP TO 375 HP @ 5200 RPM)

LOWEST-PRICED SAVOY 2-DR.
$2260. OR 2379.
(6)   (V8)

$2967.   **FURY**
CONVERTIBLE
(6948 BLT.)

118" WB (WAGONS 122")

GRILLE GUARD AVAIL. ON SOME 1961 MODELS

new GRILLE

...SOLID BEAUTY

Plymouth

AVOY

Look at Plymouth now!

WEST COAST PRICED FROM $2531.

$2609. UP

SAVOY $2262. UP

6 = 6.50 × 14 TIRES
V8 = 7.00 × 14

"PLYMOUTH" NAME ON DOOR OF SAVOY

"BELVEDERE" NAME ON DOOR OF BELVEDERE

$2342. UP

Plymouth Belvedere 2-dr Sedan
(3128 BLT.)

FROM 2563.

(ABOVE) FURY SEDAN (17,531 BLT.)

62 SP-1 (6) SP-2 (V8)

FURY (TOTALLY RESTYLED)

new 116" WB

New Forward Flair Design

FURY

FURY
4 DR. H/T
(5995 BLT.)

CONVERTIBLE
$2924.
(4349 BLT.)

$2742.

FURY TURBO-
(SPECIAL)

$2851.

Special red, white and blue insignia, new wheel covers and new rear deck design tell you that this one is the real thing! There is no mistaking a new Sport Fury—hardtop or convertible.

SPT. FURY H/T
(4039 BLT.)

NEW SPORT FURY

CONVERTIBLE (1516 BLT.)

Action! Fly to 60 mph in 8.5 secs. with optional 305-hp Golden Commando V-8 engine.

$3082.

TOTAL 1962 PRODUCTION:
177, 651

$2206. ~ 3082.
PRICE RANGE

ENGINES 225 CID 6 (145 HP)
318 CID V8 (230 OR 260 HP)
361 CID V8 (305 HP)

Plymouth

$2609. UP

SAVOY WAGON (17,216 BLT.)

BELVEDERE $2342. UP
2-DR. (6218 BLT.)

H/T (13,832 BLT.)

FURY

$2585. UP FURY

V8 OPTIONS

318 CID (230 HP @ 4400)
361 CID (265 HP @ 4400)
383 CID (320 to 330 HP)
426 CID (370 HP @ 4600 to 425 HP @ 5600 RPM)

$2924. CVT. (5221 BLT.)

1963 IS ONLY YEAR with UNUSUAL FRONT CORNER PARK./DIRECTIONAL LIGHTS

FURY 4-DR. H/T (11,887 BLT.)

MB 2560

with a 5-year or 50,000-mile warranty

7.00 x 14 TIRES

63

TP-1 (6 CYL.)
TP-2 (V8)

new GRILLE
new TAIL-LIGHTS
new FULL-LENGTH SIDE TRIM

$2742.

Get up and go Plymouth!

new 426 CID V8 ENGINE KNOWN AS "Super Stock"

TOTAL 1963 PRODUCTION: 274,735

$2206. ~ 3082.
PRICE RANGE

MOST POPULAR MODELS:
SAVOY 4 DR. SEDANS (56,313 BLT.)
BELVEDERE 4 DR. " (54,929 BLT.)
(BOTH SIXES and V8₅)

DASH

$2851.
H/T (11,483 BLT.)

A Transmission Drive Selector (optional)
B Transmission Parking Lock
C Clock (optional)
D Turn Signal Indicator
E Heater Controls (optional)
F Headlights and Panel Lights
G Defroster Outlets
H Windshield Wiper Control
I Ignition Switch
J Cigarette Lighter
K Ash Receiver
L Glove Compartment Lock
M Radio (optional)

N 1558

SPORT FURY
(3836 CVTS. ALSO)

PLYMOUTH'S ON THE MOVE

Plymouth

7.00 × 14 TIRES

$2224. UP

Savoy 2-Door Sedan

SAVOY

(21,326 BLT.)

$2620. UP   Savoy 6- or 9-Passenger Station Wagon   (15,643 BLT.)

$2444. UP   BELVEDERE   (16,934 BLT.) H/T

318, 361, 383 and 426 CID V8s   FURY

$2224. ~ 3095. PRICE RANGE

VP-1 (6 CYL.) (V8) VP-2   **64**   new GRILLE   CONVERTIBLE (5173 BLT.) $2937.

230 TO 425 HP

TOTAL 1964 PRODUCTION: 571,339 (INCL. BARRACUDA)

BELV.

$2981. UP

FURY wagon (8/28 BLT.)

$2864.

H/T (23,695 BLT.)

REAR OF FURY WAGON

$3/95. (WEST COAST)

SPORT FURY

1964

new REAR TREATMENT

new H/T ROOFLINE

new CONVEX GRILLE CLOSE UP

COWL TRIM

Plymouth

BELVEDERE II 116" WB

Belvedere Satellite

Belvedere I

SEDAN (35,968 BLT.)

$2265.

new↑
BELVEDERE
SATELLITE
←IS AVAIL. WITH
TOP-OF-LINE
426 CID V8
WITH
425 HP @
6000
RPM

BELV. has
7.35 × 14
TIRES (EXC.
WAGON)

Fury I 2 DR. (17,294 BLT.)

EU-6778

$2376.

$3209.↘

273 CID
BARRACUDA V8 ENG.
NOW AVAIL. IN BELVEDERE
(180 HP @ 4200 RPM)

(6272 BLT.)

PACE CAR AT 1965
INDY 500 RACE

Fury II

SPORT
FURY

$2478.

DASH

7.75 × 14 TIRES
8.55 × 14 (FURY WAGON

Fury III

FURY
III
H/T

65
RESTYLED

AR-1 (6-CYL.)
AR-2 (V8)

ALL FURY TYPES
GET new 119" WB
(WAGONS 121")

Fury III 4-Door Hardtop

(43,251 BLT.)

THE ROARING '65s

(21,367 BLT.)

$2863.

MORE MODEL SERIES FOR 1965 :
  BELVEDERE I, BELVEDERE II, SATELLITE ($2849. UP)
FURY I, FURY II, FURY III, SPORT FURY (119" WB)
($2376. UP) ($2478. UP) ($2684. UP) ($2960. UP)

$2226. ~ 4671. (BELV. I SUPER STOCK H/T)
BELV. I 2 DR.   (115" WB)

ENGINES :
225 CID 6 (145 HP)
273 CID V8 (180 HP)
318 CID V8 (230 HP)
361 CID V8 (265 HP)
383 CID V8 (270 OR 330 HP)
426 CID V8 (365 OR 425 H.P.)

Plymouth ...a great car by Chrysler Corporation.

Belvedere SATELLITE →

$2695.

6.95/7.35 x 14 TIRES

116" WB (117" ON WAGONS)
(BELV. / SATELL. SERIES)

(35,399 BLT.)

4/T

↑ SPORT FURY H/T
('32,523 BLT.)

$3006.
7.35/7.75 x 14 TIRES

12.1" WB ON FURY WAGONS

Satellite	*2-Door Hardtop
	*Convertible
Belvedere II	2-Door Hardtop
	Convertible
	4-Door Sedan
	2-Seat Station Wagon
	3-Seat Station Wagon
Belvedere I	2-Door Sedan
	4-Door Sedan
	2-Seat Station Wagon

BELV. SATELLITE INTERIOR

← Fury
119" WB

FURY INTERIOR

$2277.~ 3251.
PRICE RANGE

Line & Series	Body Model
VIP	*4-Door Hardtop
Sport Fury	*2-Door Hardtop
	*Convertible
Fury III	2-Door Hardtop
	*Convertible
	4-Door Sedan
	*4-Door Hardtop
	*2-Seat Station Wagon
	*3-Seat Station Wagon
Fury II	2-Door Sedan
	4-Door Sedan
	*2-Seat Station Wagon
	*3-Seat Station Wagon
Fury I	2-Door Sedan
	4-Door Sedan
	2-Seat Station Wagon

66 new GRILLES

VIP

VIP 4-DR H/T
$3133.

$3750.
(WEST COAST)

1966 PROD.:
463,971

There's something for everyone in Plymouth '66 starting with the totally new, elegant Plymouth VIP.

7 ENGINES !
225 CID SLANT-6 (145 HP)
273 CID V8 (180 HP)
318 CID V8 (230 HP)
361 CID V8 (265 HP)
383 CID V8 (325 HP)
426 CID V8 (425 HP); 440 CID V8 (365 HP)

VIP (new)

(SAME FRONT END AS FURY)

VIP DASH

4-door VIP H/T   INTERIOR

Let yourself go... Plymouth

VIP FURY BELVEDERE VALIANT BARRACUDA

The new 2-d hardtop VIP. **Plymouth** $3069.

DASH

66 ½ 2-DR. VIP $3429. (WEST)

(INTRO. 1-66

$2318.~ 3279. PRICE RANGE

DASH

7.35 x 14" TIRES $2747. (30,328 BLT.)

BELVEDERE; BELVEDERE I; BELV. II;
SATELLITE; " GTX;
FURY I; FURY II; FURY III;
SPORT FURY; VIP
MODELS AVAIL. 1967

(WEST) $3101. **Belvedere** (SATELLITE)

67 new GRILLES

(INTRO. 9-29-66)

REAR DECK DETAILS (GTX)

$3178.

Belvedere GTX (new) WITH 440 CID V8 (375 HP)

GTX H/T $3330. (WEST)

(CONT'D. NEXT PAGE)

'67 Plymouth $2872.   FURY III H/T

(37,449 BLT.)

Plymouth is out to win you over this year.
$3033.
SPORT Fury

Fury

Sport Fury
← COWL INSIGNIA

new "FAST TOP" →

SPORT FURY CVT.

FURY III $2922.

$3279.

FURY 4 DR. H/T III DETAIL

67 (CONT'D.)

$3062.

crew-size Fury wagon

(43,614 BLT.)

FURY III

$3144.

(21,803 BLT.)

REAR FACING SEAT

eng.,
drive tr.
5-year/50,000-mile
warranty

Fury

DASH

WHEEL CVR.

FURY I, FURY II, FURY III and SPORT FURY MODELS (SINCE '65)

'67 Plymouth VIP

H/T
(7912 BLT.)
$3/82.

Plymouth

(GTX has ALL HORIZONTAL GRILLE PCS.)

SATELLITE

H/T $2594.
$3047. (WEST)
(46,539 BLT.)

7.75 x 14" or 8.55 x 14" TIRES

3-SEAT $3602.

Satellite Sport Wagon

SPT. SATELLITE SERIES has 318 CID V8 (230 HP)

new CIRCULAR SIDE SAFETY LIGHTS AT EITHER END ←

(15,539 BLT.)

(INTRO. 9-14-67)

68

ROAD-RUNNER (new) with 383 CID V8 (335 HP) (426 CID V8 OPT.)

F70 x 14" TIRES

H/T $3034.
$3229. (WEST)

SPT. SUBURBAN FROM $3805. (WEST)

$3442. UP

(FURY) SPORT SUBURBAN

FURY III 4-DR. H/T $3430 (WEST)

SATELLITE WAGON (FRONT)

8.25 x 14" TIRES

(new 8-TRACK STEREO TAPE OPT.)

$3206. UP ↓

Fury

$3623. (WEST)

VIP

...the Plymouth win-you-over beat goes on ♥

1968

(BODY / FRAME WELDED INTO 1 UNIT)

SPORT FURY FROM $3569. (WEST) (2 ROOFLINES AVAIL.)

LARGEST-SELLING MODEL : **FURY III** SEDAN (57,899 BLT.) $2890.
SCARCEST PRODUCED : **GTX** CVT. (1026 BLT.) $3590.
LOWEST-PRICED : **BELVEDERE** CL. COUPE (15,702 BLT.) $2444.
HIGHEST-PRICED : **SPORT SUBURBAN** 9-PASS. WAGON (13,224 BLT.)
$3543. ($3906., WEST COAST)

1968 PROD. : 591,030

ENGINES : 225 CID SLANT-6 (145 HP);
273 CID V8 (190 HP); 318 CID V8 (230 HP);
383 CID V8 (300, 330 OR 335 HP); 426 CID V8 (425 HP);
440 CID V8 (350 OR 375 HP)

Belvedere 2509.

Plymouth CPE. $2967. (7063 BLT.) (WEST)

2883. BELVEDERE SPORT SATELLITE H/T (15,807 BLT.) (WEST) $3251.

(RESTYLED) **69** (INTRO. 9-19-68)

new AIR VANE VENTILATION FOR WAGON

Road Runner $3083. $3284. (WEST) H/T (48,549 BLT.)

$3303. FORMAL H/T

SPORT FURY $3671. (WEST)

Sport Suburban (OTHER VIEW ABOVE)

A completely new Fury for 1969.

$3718.

$4086. (WEST) (3-SEAT)

STARTING 1969, PLYMOUTH SIDE LIGHTS ARE RECTANGULAR.

DASH (FURY)

VIP GRILLE

VIP FROM $3382.

Look what Plymouth's up to now:

$2509.~3718. PRICE RANGE

MODELS: RL BELVEDERE ; RH SATELLITE ; RP SPT. SATELLITE ; RM ROAD RUNNER ; RS GTX ; PE FURY I ; PL FURY II ; PM FURY III ; PH SPORT FURY ; PP VIP ; EP SUBURBAN WAGONS

1969 PRODUCTION: 581,004

Pontiac

MFD. BY PONTIAC DIVISION OF GENERAL MOTORS CORP.

(SINCE 1926)

$1307. ~ 2047. PRICE RANGE

Finest of the Famous "Silver Streaks"

TOTAL 1946 PROD.: 137,640

$1631. UP

TORPEDO

239.2 CID 6 (90 HP)
248.9 CID STRAIGHT-8 (103 HP)

STREAMLINER SEDANS ARE FASTBACKS

STREAMLINER

WAGON $2019. UP

46 new GRILLE

$1510. (6)
1538. (8)

WHAT'S NEW AND IMPROVED IN THE 1946 PONTIAC

New, beautiful exterior appearance . . . New instrument panel . . . Heavier chrome finish . . . Improved, rust-resistant bodies . . . New interior trim . . . Improved clutch . . . New, wider wheel rims . . . Longer-life muffler and tail pipe . . . Improved cooling.

STREAMLINER

$1387. ~ 2359. PRICE RANGE

TORPEDO
$1512. UP

2 DR. 1547. UP

47 NO VERTICAL PIECES IN 1947 GRILLE

TOTAL 1947 PROD.: 230,600
(OR 223,015)

TORPEDO SPT. CPE.

TORPEDO DELUXE

$1552. UP

$1731. UP

48 new GRILLE

$2442. UP

$1817. UP

STREAMLINER DE LUXE

DE LUXE MODELS *have* CHROME STRIP ON SIDE OF FRONT FENDER, *and* CHROME REAR FENDER PADS

$1766. UP

TORPEDO DELUXE CVT.
$2025.

HYDRA-MATIC A/T NOW AVAILABLE

new 104 HP IN STRAIGHT-8

$1500. ~ 2490. PRICE RANGE

TOTAL 1948 PRODUCTION: 235,419
(OR 253,469)

PONTIAC

CHIEFTAN CVT.
$2138. UP →

STREAMLINER

STREAMLINER (FASTBACK)

TOTAL 1949 PROD.
333,957

$1740. (6)
1808. (8)

CHOICE OF METAL OR WOOD-BODY WAGONS IN 1949
$2622. (6) $2690. (8)

49

(TOTALLY RESTYLED)

models
STREAMLINER 6 OR 8
CHIEFTAN 6 OR 8 (REPLACES "TORPEDO")
ENGINES = 239.2 CID 6 (90 OR 93 HP); 248.9 CID STRAIGHT-8 (104 OR 106 HP)
7.10 × 15" TIRES

120" W.B. (THROUGH '52)

CHIEFTAN
SEDAN $1761. (6)
1829. (8)

90 HP 6 OR
108 HP (268.4 CID) 8

TOTAL 1950 PRODUCTION: 467,655 (OR 446,429)

$1673. ~ 2411. PRICE RANGE

STREAMLINER

$2343. UP

DELUXE ALL-STEEL WAGON

CHIEFTAN
CLUB COUPE
$1694. UP

CHIEFTAN

50

VERTICAL "TEETH" NOW ADDED TO UPPER SECTION OF GRILLE

SUPER CATALINA H/T (new)
$2058. UP

WHEEL COVER

CHIEFTAN
SEDAN
$1745.

CHIEFTAN (RUBBER REAR FENDER GUARDS IDENTIFY STD. MODELS)
STD. $1977. DLX. $2081.
(8)

WAGON

STREAMLINER DLX. $2629.
(8)

DLX. SEDAN

"Pontiac Eight" ABOVE TRIM

REAR FENDER PAD DETAILS (CHROMED on DELUXE)

$1713.~2629. PRICE RANGE

TOTAL 1951 PRODUCTION: 343,795

TAILLIGHTS SLIGHTLY ENLARGED

51 new GRILLE

new 102 HP (6) or 122 HP (8)

L-HEAD 6 CYL. and STRAIGHT-8 ENGINES CONTINUE (THROUGH '54)

('52) LESS SIDE CHROME on STANDARD MODELS.

Dollar for Dollar you can't beat a **Pontiac**

new 1952 WHEEL COVER

52

4 new "SLOTS" BELOW PONTIAC NAME (NOW IN CAPITALS)

SIDE VIEW OF MASCOT FOR 1952

CHIEFTAN

PONTIAC NAME OVER GRILLE IS IN Cursive SCRIPT AS PREVIOUSLY

CVT. $2388.
(8)

New High-Performance Economy Axle

More Power

New Dual-Range Hydra-Matic Drive*

(IMPROVED AUTOMATIC TRANSMISSION)

new SIDE TRIM

TOTAL 1952 PRODUCTION: 277,156

(SAME ENGINES and HP AS IN '51)

CHIEFTAN MODELS ONLY (THROUGH '53)

$1956.~2772. PRICE RANGE

330

PONTIAC
CHIEFTAN DE LUXE
$2119. UP

LOWEST PRICED MODEL, AT
$1956.

CHIEFTAN SPECIAL

$2060. UP

new "DUAL STREAK" RESTYLING
FEATURES TWIN GROUPS of
CHROME BANDS ALONG HOOD
and DECK, with new BODIES

CHIEFTAN DE LUXE

53 new 1-PIECE WINDSHIELD

TOTAL 1953 PRODUCTION:
414,011

new 122" WB

$1956.

$2774.

PRICE RANGE

WAGONS with GRAIN-DECORATED
UPPER PANELS (ABOVE) ARE PRICED
$80. ABOVE SIMILAR WAGONS
of ONE SOLID COLOR ONLY.

WOODGRAINED
DLX. WAGON
(8) IS
COSTLIEST
MODEL
AT
2744.

ENGINES:
115 or 118 HP 6
118 or 122 HP STRAIGHT 8

The experimental Parisienne stands only
56 inches high. Inside and out it is a
designer's dream of how one "car of the
future" might be styled and equipped.

PARISIENNE

(SHOW CAR)
PUBLICLY DISPLAYED, BUT
NOT A PRODUCTION BODY
TYPE

De Luxe Catalina

CHIEFTAN DLX.
SEDAN
$2194.

DELUXE CATALINA 6 H/T $2304.
CUSTOM   "   "   "   2370.
DELUXE   "   8   "   2380.
CUSTOM   "   "   "   2446.

DOLLAR FOR DOLLAR YOU CAN'T
beat a
Pontiac

PONTIAC

122" WB

CHIEFTAN SPECIAL 6
(ALSO AVAIL. as 8)

$2027.
(8 CYL., 2102.)

$1968.~2630.
PRICE RANGE

ENGINES:
239.2 CID 6 (115 OR 118 HP)
268.4 CID STRAIGHT-8 (122 OR 127 HP)

CHIEFTAN DELUXE
(6 OR 8 CYL.)
2 DR. 6 CYL. SEDAN
$2131.

A BRIEF (1 YR.) RETURN TO SINGLE GROUP OF CHROME STRIPS ON HOOD and DECK

new GRILLE with EMBLEM PLACED ABOVE

54

6-Z or 8-Z

DELUXE (ABOVE) HAS MORE SIDE TRIM THAN SPECIAL (TOP)

WAGON INTERIOR

The rear seat in this handsome 6-passenger interior, in green or red Morrokide with ivory trim, folds forward to provide almost six feet of cargo space—eight feet with tail gate open

122" WB

CHIEFTAN DELUXE STATION WAGON

STAR CHIEF DE LUXE 8 (BELOW) COSTS LESS THAN CUSTOM 8.

CONVERTIBLE $2630.

$2504.(6); 2579.(8)

124" WB

Dollar for Dollar
You Can't Beat a
PONTIAC

STAR CHIEF CUSTOM 8
(AT RIGHT and BELOW)
(115,088 STAR CHIEF DLX. and CUSTOM MODELS BLT.)

$2394.

THE NEW *Star Chief*

8-CYL. 124" WB
SERIES "8-Z"
IDENTIFIED BY THESE "STARS" ON SIDE OF REAR FENDER

TOTAL 1954 PRODUCTION:
370,887 (OR 287,744, DEPENDING ON VARYING REPORTS)

PONTIAC
DR. (58,654 BLT.)

CHIEFTAN 860

$2105.

(TOTALLY RESTYLED)

55

4 DR. WAGON (19,439 BLT.)
$2603.

new PANORAMIC WINDSHIELD

2 or 4 DR. 860 WAGONS

Pontiac leads in station wagon value with four models —the beautiful 860, left, in two- and four-door models, the spectacular 870 four-door and the fabulous Safari.

CHIEFTAN 870

$2105.
TO $2962.
PRICE RANGE

(91,187 BLT.)
$2268.

CHIEFTAN = 122" WB
STAR CHIEF = 124" WB

new DASH

CHIEFTAN 870
CATALINA H/T
(2 VIEWS)

STAR CHIEF
CONVERTIBLE
(19,762 BLT.)

$2335.

(72,608 BLT.)

STAR CHIEF CUSTOM
CATALINA

$2691.

ALL MODELS with new "STRATO-STREAK" O.H.V. V8 ENGINE →
180 OR 200 HP
(287.2 CID)

$2499.
(99,629 BLT.)

FRONT DETAILS CLOSE UP

new STAR CHIEF CUST. SAFARI 2 DR. LUXURY WAGON

Pontiac's flair for years-ahead styling was never more evident than in the fabulous all-new Safari.

3760 BLT.)
$2962.
SAFARI SIMILAR TO CHEVROLET NOMAD WAGON)

REAR 3/4 VIEW OF CHIEFTAN 870 STATION WAGON

TOTAL 1955 PRODUCTION : 581,860

new 12-VOLT ELECTRICAL SYSTEM

BUMPER GUARDS AVAILABLE (RARE)

2 DR. WAGON

CHIEFTAN

(8620 BLT.)

SAFARI

(4042 BLT.)

STAR CHIEF

new
4-DR. H/T

CHIEFTAN 860
8-PASS.
WAGON
$2613.

(12,702
BLT.)

56 Pontiac
Star Chief
Custom Convertible

PLASTIC
"JR. STAR CHIEF"
CHILD'S ELECTRIC
CARS ALSO, FOR
DEALER PROMOTIONAL
PURPOSES

(19,762
BLT.)

$2691.

NEW GRILLE WITH
MORE CHROME

REAR
DETAILS

ENLARGED
new 316.6 cid V8
(205 or 227 HP)

$2240. ~ 3129.
PRICE RANGE

TOTAL 1956 PRODUCTION:
332, 268

PONTIAC

CHIEFTAN 252 HP

$2527. (35,671 BLT.)

new 4-DR. "SAFARI" (14,095 BLT.)

SUPER CHIEF 270 HP

$3021.

8.50 x 14 TIRES ON WAGONS

$3481.

57 RESTYLED

SU. CHIEF WAGON DOOR

STAR CHIEF 270 HP CUSTOM CATALINA 4-DR. H/T

CUSTOM SAFARI WAGON (1292 BLT.)

$2975.

(44,283 BLT.)

$5782.

BONNEVILLE (new)

(630 BLT.)

(LIMITED PRODUCTION)

$3105. 72,789 BLT.

STAR CHIEF has HEAVY CHROME BAND PLACED WITHIN COLOR CONTRAST PANEL ON REAR FENDER

AMERICA'S NUMBER ① ROAD CAR!

new FRONT END (NO LONGER USES "Silver Streak" CHROME BANDS)

new 8.00 x 14" TIRES (7.50 x 14 CHIEFTAN)

STAR CH. 2-DR. H/T (CUSTOM CATALINA) (32,864 BLT.)

$2901.

ENGINES:
347 CID V8
(252, 270 or 290 HP)
370 CID (BONNEVILLE)
(300 HP w. FUEL INJECTION)

IN 1957, ALL PONTIAC STATION WAGONS (2 OR 4 DR.) CARRIED THE "SAFARI" NAME. BUT ONLY THE 2-DR. CUSTOM SAFARI (1955~1957) IS THE "CLASSIC" MODEL SO DESIRED BY COLLECTORS.

$2463. ~ 5782. PRICE RANGE

TOTAL 1957 PRODUCTION: 343,298

PONTIAC

CHIEFTAN

DASH

CATALINA 4-DR. H/T (17,946 BLT.) $2792

3 STAR-LIKE FIGS. ON REAR FENDER OF CHIEFTAN; 4 ON SUPER CHIEF. (EACH MODEL CAN BE IDENTIFIED BY FENDER DECOR., AS ILLUS. BELOW)

SUPER CHIEF

370 CID V8 (240, 255, 270, 285, 300 OR 310 HP)

58 (RESTYLED)

H/T (7236 BLT.) $2880.

STAR CHIEF H/T

BOLDEST ADVANCE IN 50 YEARS

BOLD NEW Bonneville BY PONTIAC

$3586. (3096 BLT.)

PACE CAR AT 1958 INDY 500 RACE

3122.

(13,888 BLT.)

$3481.

CVT.

BONNEVILLE H/T

(9144 BLT.)

models

MODEL MOST PRODUCED = CHIEFTAN 4 DR. (44,999)
LEAST = STAR CHIEF CUST. SAFARI WAG. (2905)

models		PRICE RANGE	
CHIEFTAN 122" WB	$2573. ~ 3088.		
SUPER CHIEF 124" WB	2834. ~ 2961.	"	"
STAR CHIEF (CUSTOM) 124" WB	3071. ~ 3350.	"	"
BONNEVILLE (CUSTOM) 122" WB	3481. (H/T); 3586. (CVT.)		

TOTAL 1958 PRODUCTION: 219,823

PONTIAC

CATALINA
SEDAN (72,377 BLT.)
$2702.

new "WIDE-TRACK"
59
(TOTALLY RESTYLED)

122" WB (CATALINA SERIES and ON BONNEVL. WAGON)
124" WB ON OTHERS

$3333.

new BONNEVILLE VISTA (38,696 BLT.) 4-DR. H/T

"BONNEVILLE" NAME ON BONNEVL. GRILLE (ABOVE)

CATALINA SAFARI WAGON

CATALINA WAGON (14,084 BLT.) (9 PASS.) (21,162 6 PASS.)

BONNEVILLE H/T (27,769 BLT.)

3207. (9-PASS.)

(WIDTH EXAGGERATED)

"PONTIAC" NAME ON GRILLE OF CATALINA and STAR CHIEF. STAR CHIEF has STAR-LIKE FIGURES ALONG SIDE OF REAR FENDER.

DASH

8.00 × 14 TIRES

$3257.
245 HP (280 w. Hydra Matic)
BONNEVL. has 260 HP (300 w. Hyd.)

new 389 C.I.D. V8 ENGINE (IN ALL MODELS)
(215, 245, 260, 280, 300 OR 310 HP)

models		
CATALINA	122" WB	$2633. ~ 3209.
STAR CHIEF	124" WB	2934. ~ 3138.
BONNEVILLE (and "CUSTOM SAFARI")		3257. ~ 3532.
124" WB	122" WB	

TOTAL 1959 PRODUCTION:
388,856

337

PONTIAC 1960

TOTAL 1960 PRODUCTION : 418,154

STAR CHIEF SEDAN (23,038 BLT.) THE ONLY CAR WITH WIDE·TRACK WHEELS $3003.

WAGON AND VISTA DETAILS

CVT. (17,062 BLT.) $3476. BONNEVILLE

$2971.

60

new GRILLE WITH HORIZONTAL PCS.

H/T

VENTURA H/T (new) (27,577 BLT.)

389 CID V8 (215, 245, 260, 283, 303 OR 318 HP)

(24,015 BLT.) $3255.

389 CID V8 (215, 230, 235, 267, 283, 287, 303 OR 318 HP)

61 (RESTYLED)

PRODUCTION : (FULL-SIZED) 244,391

$2631. ~ 3530. PRICE RANGE

new PONTIAC TEMPEST COMPACT SERIES (SHOWN SEPARATELY)

$2631.

CATALINA

2-DR. SPT. SEDAN

(9846 BLT.)

SAFARI

models
CATALINA
VENTURA
STAR CHIEF
BONNEVILLE

BONNEVILLE H/T (STAR CHF. TAILLIGHTS SIMILAR) (16,906 BLT.) $3255.

(CONT'D. NEXT PAGE)

TOTAL 1961 PRODUCTION : 244,391

PONTIAC

STAR CIHEF VISTA
4 DR. H/T
(13,559 BLT.)
3/36.

61
CONT'D.

BONNEVILLE
OFTEN *has*
"BONNEVILLE"
NAME ON GRILLE
AS ILLUSTR.

VISTA 4 DOOR HARDTOPS

$3230.

STAR CHIEF (13,882 BLT.)

STAR CHIEF VISTA

VISTA
4 DR. H/T

$3331.

(30,830 BLT.)

Wide-Track Pontiac
WIDEST STANCE ON THE ROAD

62
new
GRILLES
and TAIL LIGHTS

62-III

SPORT
H/T
(31,629 BLT.)
$3349.

Bonneville

CATALINA
9-PASS. SAFARI WAG.
$3301.

(10,716 BLT.)

CATALINA 9-PASSENGER SAFARI

62-100

CLOSE-UP
OF
FRONT
END

389 CID V8
(215,230,235,267,283,
303,318,333 OR 345 HP)

(CONT'D. NEXT PAGE)

1962 PRODUCTION: 401,674

339

STRATO - CHIEF
(SOLD ONLY IN CANADA)

WAGONS

LAURENTIAN
2 - DOOR

PONTIAC

LAURENTIAN
(SOLD ONLY IN CANADA)

62
(CONT'D.)

SPORT COUPE

SPORT COUPES

PARISIENNE
(SOLD ONLY IN CANADA)

SEDAN

GRAND PRIX (REAR)

GP

AMERICAN
GRAND PRIX
New H/T →

MANUAL
SHIFT
CONSOLE

BUCKET
SEATS

AUTO.
SHIFT
CONSOLE

TACH.

$3490.
303 HP

4 BBL. CARB.

$3917. (WEST COAST)

GRAND PRIX (GP) DETAILS (ABOVE)   2-DR. GP H/T ONLY   (30,195 BLT.)

$2725. ~ 3624. PRICE RANGE

PONTIAC $3300.

CATALINA

9-PASSENGER SAFARI

(11,751 BLT.)

$2795.
CATALINA SEDAN
(79,961 BLT.)

$3096.

(28,309 BLT.)

STAR CHIEF 4-DOOR SEDAN

← BEST-SELLING 1963 PONTIAC

(30,955 BLT.)

$3348.

(14,091 BLT.)

$2725.
CATALINA SPORTS SEDAN

BONNEVILLE SPORTS COUPE

BONNEVILLE VISTA

$3423.

(49,929 BLT.)

CATALINA / ST. CHIEF

CATALINA

SPORT H/T
(60,795 BLT.)
$2859.

ENGINES (V8s)
389 CID (215, 230, 235, 267, 283, 303 OR 318 HP)
new 421 CID (353 OR 370 HP)

BONNEVILLE

CVT. (23,459 BLT.)

$3568.

63 new GRILLE

$2725. ~ 3623.
PRICE RANGE

2 VIEWS OF G.P.

GP
PONTIAC GRAND PRIX

SPORT H/T
(72,959 BLT.)

$3489.

'63 WIDE-TRACK PONTIAC

TOTAL 1963 PRODUCTION:
481,652

341

PONTIAC 120" WB

(12,480 BLT.)

$2735.

CATALINA

STAR CHIEF
(FRONT SIMILAR TO CATALINA)

123" WB
235 HP

$3203. UP

64 new GRILLES

(26,453 BLT.)

$3107.

$2735. ~ 3633. PRICE RANGE

$3433.

4-DR.
H/T
(57,630
BLT.)

CATALINA SEDAN

(84,457 BLT.)

$2806.

$3578

BONNEVILLE

BONNEVILLE

WHEEL COVER

123" WB

BONN. BROUGHAM
W. VINYL TOP (BELOW)

INTERIOR

WEST COAST

$3995.
(CVT.)
(22,016 BLT.)

306 HP

PRESTON CLOTH-and-MORROKIDE INTERIOR

G.P.

WAGON
(BONNEVILLE)

SAFARI 4 DR.   (5844 BLT.)
$3633.

ENGINES : 389 CID (230 TO 330 HP)
421 CID (350 OR 370 CID

120" WB
(63,810 BLT.)

$3499.

TOTAL 1964 PRODUCTION : 443,306

342

PONTIAC

CATALINA
121" WB

CATALINA 2+2

note LOUVRES ON COWL OF 2+2

$3594.

(21,050 BLT.)

CVT.

BONNEVILLE

BON. BROUGHAM INTERIOR

BONNEVILLE (325 HP)
BROUGHAM 4-DR. H/T

DASH →

ENGINES:
389 CID V8
(256, 290,
325 OR 333 HP)
421 CID V8
(338, 356
OR 376 HP)

$2734.~3632.
PRICE RANGE

TOTAL 1965
PRODUCTION:
534,633
(FULL-SIZED)

(62,480 BLT.)

$3433.

new GRILLES and PROTRUDING CENTER "NOSE"

$3433.

65

(57,881 BLT.)
$3498.

G.P.

CAT. 2+2 INTERIOR

Pontiac for 1965
The year of the Quick Wide-Tracks

new GRILLE

Grand Prix

GM MARK OF EXCELLENCE **PONTIAC**

full-sized cars

V-8 ENGINES (SINCE 1955)

CATALINA COWL LETTERING IN SCRIPT; OTHER MODELS WITH NAMES IN BLOCK LETTERS.

$32/9. CVT. (14,837 BLT.)

$3628. (WEST)

252 Catalina

389 CID V8 (290 HP) 121" WB

← $3240.

$3217. UP

ENGINES (V8s)

389 CID (256, 290, 32.5, OR 333 HP)

421 CID (338, 356 OR 376 HP)

66

new GRILLES

$3602.

$4011. (WEST)

254 **2+2**

421 CID V8 (338 HP) (356 OR 376 HP AVAIL.) 121" WB

Wide-Track Pontiac/'66

"2+2" LETTERING

CATALINA WAGONS FROM $4164. (WEST)

DASH

1966 PRODUCTION : 481,591

Ventura

256 **Star Chief Executive**

389 CID V8 (290 HP)

124" WB

REAR FENDER TRIM

H/T (new) $3590. (WEST)

$3/70. (BASE PRICE)

INTERIOR (EXEC.)

H/T (10,140 BLT.)

(CONT'D. NEXT PAGE)

344

PONTIAC

$4385.

WAGONS : 121" WB

Wide-Track Pontiac/'66

"BONNEVILLE" NAME ALSO SEEN ON GRILLE OF

262 Bonneville

124" WB

BONNEVILLE WAGON $4704.

$4543. (TWEST)

STAINLESS STEEL LOWER SIDE TRIM (BNVL., GP)

BONNEVILLE CVT. (16,299 BLT.) $3586. (BASE PRICE)

(8452 BLT.) $3747. (BASE PRICE)

BONNEVILLE BROUGHAM OPTION

$2762.~3747. PRICE RANGE

66 (CONT'D.)

389 C/D V8 IN BON., GP (333 HP)

GRAND PRIX H/T (36,757 BLT.) $3492.

GRAND PRIX has OWN REAR STYLING, 121" WB

266 Grand Prix

$4449.

AVAIL. ONLY AS A 2-DR. H/T

GRAND PRIX has "GP" and RALLY LTS. ON GRILLE

GRAND PRIX has OWN GRILLE

2-DR. H/Ts ONLY, IN GP SERIES, UNTIL 1967.)

SIDE DETAILS

UP TO 376 HP AVAIL. (GP)

WOOD-GRAIN ON GRAND PRIX DASH

Pontiac/67

new 400 CID V8 (290 HP) (265 325 333 and 350 HP ALSO)

new 8-TRACK STEREO TAPE SYSTEM OPTIONAL

new 428 CID V8 ALSO (360 or 376 HP)

Where did they hide the windshield wipers on the 1967 Pontiacs?

Only your Pontiac dealer knows. And he's not talking until September 29.

Catalina

CVT. (10,033 BLT.)

$3276.

TOTAL 1967 FULL-SIZE PRODUCTION: 445,950

"PONTIAC" NAME ON GRILLE

1967 CA

(WEST) $3274.
$2866. BASE PRICE

(5903 - 6 PASS.) (5593 BLT., 9 PASS.)
$3600. UP

*=new WIPERS CONCEALED BELOW REAR END OF HOOD

(new) EXECUTIVE SAFARI WAGON

new ENERGY-ABSORBING STEERING COLUMN; new 4-WAY HAZARD FLASHER, DUAL MASTER CYLINDER BRAKE SYSTEM

BONNEVILLE H/T 400 CID V8 (325 HP)

FR. $4557.

new GRILLES

67

8.55 × 14 TIRES

UP TO 376 HP WITH new 400, 428 CID V8s.

Bonneville

BONNEVILLE H/T (REAR QUARTERS)

(INTRO. 9-29-66)

(31,016 BLT.) "BONNEVILLE" NAME ABOVE GRILLE

$3448
$4405 (WEST)

new FRONT DISC BRAKE OPTION

'67 Grand Prix (GP)

(5856 CVTS. BLT.)

GP $4770. CONVERTIBLE ADDED TO LINE (1967 ONLY)

GP

new CONCEALED HEADLIGHTS (ONLY ON GRAND PRIX)

(37,125 GP H/Ts ALSO BLT.)

GP 400 CID V8 has 350 HP

67-108

Pontiac 67/Ride the Wide-Track Winning Streak

PONTIAC

CATALINA, VENTURA, EXECUTIVE
400 CID V8 STD.
(265 OR 290 HP)
(428 CID V8 OPT.)

$3744. UP

FROM $4723. (WEST)

CATALINA
H/T (92,217 BLT.)
$3518. (WEST)

METAL "NOSE" MORE PRONOUNCED THAN BEFORE.

(INTRO. 9-21-67)

68

EXECUTIVE SAFARI WAGON

"MORROKIDE" (IMITATION LEATHER) and CLOTH UPHOLSTERY AVAIL.

new GRILLES, TAIL LTS.

3089. (BASE PRICE)

VENTURA
4-DR. H/T

(41,727 BLT.)
(CATALINA SERIES, VENTURA PACKAGE OPTIONAL)

$3158. w/o VENTURA OPTION

$3592.

$4571. (WEST)

DASH (BONNEVILLE BROUGHAM)

400 CID V8 (340 HP)

124" WB BONNEVILLE
H/T (WITH TAIL LIGHT DETAILS ABOVE)
(57,055 BLT.)

8.55 x 14" TIRES STD.

$3697. (31,701 BLT.)

GP
400 CID
V8
(350 HP)

GRAND PRIX H/T
$4676. (WEST)

WAGONS have
400 CID V8
(265, 290 OR 340 HP)

(GP CVT. DISCONTINUED)

$2945. ~ 3987.
PRICE RANGE

ENGINES: 400 CID V8 (265, 290, 340, OR 350 HP)
428 CID V8 (375 OR 390 HP)

TOTAL 1968 FULL-SIZE PRODUCTION:
484,849 (910,482 = ALL SIZES)

347

Pontiac announces the great break away!

$3519. UP

Pontiac Station Wagons

$4104.

$4490. UP (WEST) CATALINA

$5081. (WEST) BONNEVILLE

EXECUTIVE SAFARI

$3872. UP

FROM $4849. (WEST)

CATALINA, VENTURA, EXECUTIVE have THIS GRILLE

new GRILLES 69 (INTRO. 9-26-68)

TYPICAL PONTIAC MEDALLION

RECTANGULAR-CLUSTER NON-GP DASH

4 DR. H/T $3756.

BONNEVILLE $4733. (WEST)

428 CID V8 (360 HP) BONNEVILLE

new 125" WB (50,817 BLT.)

GRAND PRIX

new 118" WB

GP NOW has UN-SHROUDED INDIVIDUALLY-SET HEADLIGHTS

GP new "WRAP-AROUND" DASH $3866.

$4853. (WEST)

400 CID V8 (350 HP)

(112,486 BLT.)

GP IS BEST-SELLING '69 PONTIAC

400 CID V8 USED IN MOST FULL-SIZED 1969 PONTIACS, EXCEPT BONNEVILLES.

$3090. ~ 4104. PRICE RANGE

TOTAL 1969 PRODUCTION, FULL-SIZE: 493,453 (870,081=ALL SIZES)

PUP MOTOR CAR CO., SPENCER, WIS.

PUP

(1947-1949)

50~60 AVG. MPG.

TOP SPEED 35~40 M.P.H.

1 or 2 CYL., 7½ OR 10 HP

48-49

(WOODEN BODIES)

$500.

(PRICES APPROX.)

REAR ENGINE

$600.

AT NASH DEALERS

Rambler 6 CYL. (172.6 CID)

(1950-1969)

TOP DOWN
TOP UP

$1808., f.o.b.

(CVT. OR WAGON)

100" WB

all new

50

ENGINE: 172.6 CID 6 (82 HP) (THROUGH '52)

MODEL 5021
New Rambler Convertible Landau
(CUSTOM SERIES)

"RAMBLER" NAME REVIVED BY NASH FOR THIS NEW COMPACT SERIES. CVT. INTRO. 3-50; WAGON 5-50

5.90 × 15

51-52

new "SUPER"

$1968. ('51) 2094. ('52)

MODEL 5127 (5227)

"COUNTRY CLUB" HARDTOP INTRO. 6-51

SUBURBAN MODEL 5114 (5214)

('51) $1885.
('52) 2003.

57,555 RAMBLERS BLT. 1951; 53,055 RAMBLERS SOLD IN 1952

FIRST 2 DIGITS OF MODEL NUMBER INDICATE YEAR OF CAR

CONVERTIBLE CONTINUES $2119. ('52)

EST. PRODUCTION TOTAL: 1950 (21,674)

RAMBLER

2150.

CUSTOM

5321 CVT.

5324 2-DR. WAGON

85 HP (90 *with* Hydra-Matic)

53

$2125.

5327 COUNTRY CLUB H/T

GREENBRIER

SUPER 2-DR. SUBURBAN IS LOWEST-PRICED : **$2003.**, f.o.b.

ENGINES : *new* 184 cid 6 (85 HP)
new 195.6 cid 6 (95 HP w. AUTO TRANS.)

TOTAL 1953 PRODUCTION : 41,825

$2003.~2150. PRICE RANGE

5406 CLUB SEDAN

new DE LUXE

NASH and HUDSON MERGE TO CREATE AMC (AMERICAN MOTORS CORP.) MAY 1, 1954

SUPER

$1800. 5417

$1550.

$1800.

5414 SUBURBAN WAGON

COUNTRY CLUB H/T

Nash Motors, Division of AMERICAN MOTORS CORP.
DETROIT, MICH.

54

100" and 108" WB

$1965.
5425 CUSTOM SEDAN

$1980.

5421 CONVT.

← CUSTOM

note LUGGAGE RACK AND DIP IN REAR ROOFLINE

TOTAL 1954 PROD. : 37,779

REAR DETAILS (CROSS-COUNTRY)

new 4-DR. "CROSS-COUNTRY" WAGON (5428)

$2050.

RAMBLER a Whole **New Idea** in Automobiles

$1869.

5514

DELUXE

5515

SUPER SUBURBAN WAGON

PRICES START AT

$1585.,
f.o.b.
(DLX. 2-DR.)

NEW IDEA! *Touch this knob—and it will always be springtime in your Rambler. No cold in winter! No heat in summer! No dust or traffic roar! You breathe only fresh, filtered air. It's American Motors' All-Season Air Conditioning*— greatest health, comfort, safety feature of fifty years. Needs no trunk space. And you buy a Rambler so equipped for less than the price of an ordinary car!*
Patents applied for

$1995.

5517-2
COUNTRY
CLUB
CUSTOM H/T

NEW GRILLE with CRISS-CROSS PIECES

$1995.

ON 9-22-55, FINAL AMC CAR ASSEMBLED AT EL SEGUNDO, CALIF. BRANCH FACTORY, with "DC-" SERIAL NUMBERS. KENOSHA, WIS. FACTORY CONTINUES with "D-" SERIAL NUMBERS AS USUAL.

55

"FLEET" BUDGET MODEL 2-DR. SEDAN, 2-DR. WAGON and 4-DR. WAGON AVAIL.

195.6 CID 6 [90 OR 100 HP

TOTAL 1955 PRODUCTION: 83,852

$2096.

$1585.~ 2098.
PRICE RANGE

INTERIOR (H/T)

CROSS-COUNTRY

5518

1955½, WITH "SPRING SPECIAL" TRIM

HUDSON RAMBLER

NASH RAMBLER
(DETERMINE BY NAME-PLATE ON GRILLE)

NOW AT *Nash* DEALERS AND **HUDSON** DEALERS EVERYWHERE

American Motors

RAMBLER
You'll make the Smart Switch for '56

Product of American Motors

AMERICAN MOTORS MEANS [logo] MORE FOR AMERICANS

See Disneyland—great TV for all the family over ABC network.

DE LUXE 5615 — $1829.

SUPER 5618-1

$2230 — America's lowest-priced 4-door station wagon, delivered at the factory, including federal taxes. State and local taxes (if any), white wall tires and optional equipment (if desired), extra.

YOU SAVE ON FIRST COST. Model for model, Rambler is lowest-priced of all, with similar equipment, yet you get luxuries that rival the $5,000 cars—Power Brakes standard on custom models!

YOU SAVE ⅓ ON GASOLINE. New Typhoon OHV engine, with 33% more power, delivers up to 200 more miles on a tankful than other low-price cars.

new 120-HP SIX with OVERHEAD VALVES

new BROADER GRILLE ENCOMPASSES HEADLIGHTS

56
(RESTYLED)
108" WB ON ALL

Box-Girders All Around Passengers — Box-Sections Absorb Impact — The Old Way — SEE THE DIFFERENCE

Make the Smart Switch to Double Safe Single Unit Car Construction. All-welded, twice as rigid with "double lifetime" durability—means higher resale value.

NOTE VARIATIONS IN UPPER SIDE TRIM

CUSTOM SEDAN 5615-2

$2059.

Make the Smart Switch to the car that out-corners, out-parks them all. Entirely new ride—first low-priced car with Deep Coil Springs on all four wheels.

$1829. ~ 2494.
PRICE RANGE

TOTAL 1956 PRODUCTION: 79,162
(OR 79,166)

"Make the Smart Switch to Rambler!"

$2224.

new COLOR SPEAR SIDE MOULDINGS ON CUSTOM

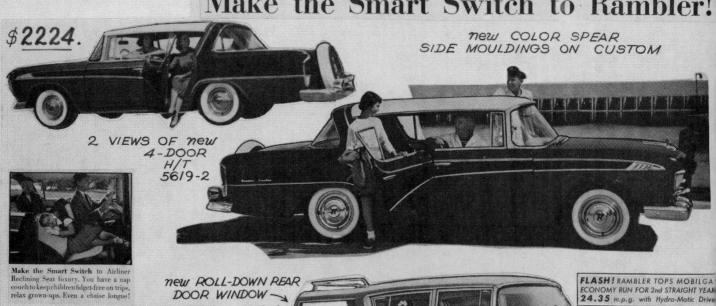

2 VIEWS OF new 4-DOOR H/T 5619-2

Make the Smart Switch to Airliner Reclining Seat luxury. You have a nap couch to keep children fidget-free on trips, relax grown-ups. Even a chaise longue!

new ROLL-DOWN REAR DOOR WINDOW 5618-2

New Rambler Cross Country Station Wagon! Enjoy more fun per mile and per dollar in America's lowest-priced four-door Custom Station Wagon.

CUSTOM

$2329.

FLASH! RAMBLER TOPS MOBILGAS ECONOMY RUN FOR 2nd STRAIGHT YEAR 24.35 m.p.g. with Hydra-Matic Drive.

OVERDRIVE AVAIL.

RAMBLER

(6) 5715-2
(V8) 5725-2

CUSTOM

new 190 H.P. V·8
And Economy 6

5718-1 (6)
5728-1 (V8)

$2213. (6)
2343. (V8)

(6) $2410. →
(V8) $2540.

SUPER

5718-2 (6)
5728-2 (V8)

CROSS-COUNTRY
WAGON

GEORGE ROMNEY
PRESIDENT, AMERICAN MOTORS
(UNTIL 2-62)

$2500. (6)
2630. (V8)

new REBEL V8

5739-2

255 HP
327 CID V8

← CUSTOM

AVAIL. AS REGULAR
OR HARDTOP
WAGON ↘

(1500 BLT.)

CUSTOM REBEL 4-DR. H/T $2786.

$1961.~2786. PRICE RANGE

2715.

5723-2 (V8)

ENGINES :
195.6 CID 6 (135 HP)
new 250 CID V8 (190 HP)
new 327 CID V8 (255 HP)

$2428.

SIDE MOULDINGS
CHANGED

Rambler Custom

6.70 x 15

57 20 MODELS

SMALL EXTRA PIECE
ADDED, IN TOP SECTION
OF GRILLE

5729-2 (V8)

114,084
RAMBLERS BLT.
CALENDAR YEAR, 1957
(91,469 MODEL YEAR)

RAMBLER

5802 or 5806

5815

DE LUXE 6
$2047.

new AMERICAN 6
(100" WB)
42,196 SOLD

$1775., f.o.b.
and up

127 HP 6
SUPER
5815-1

58

new GRILLES

new TAIL-FINS
(ON ALL BUT
AMERICAN)

CUSTOM 6

MORE THAN 100
IMPROVEMENTS!
22 MODELS

INTRODUCED FOR 1958,
new AMERICAN and AMBASSADOR
MODELS have
OWN GRILLES, DIFFERENT
FROM THOSE OF
OTHER RAMBLERS.

$2532. →

215-HP
REBEL V8

REAR DETAILS

REBEL V8
CUSTOM COUNTRY CLUB
4 DR. H/T

5829-2

DASH

5888-1 or 2

new AMBASSADOR
V8 117" WB
new = 4
HEADLIGHTS 270 HP
(THROUGH '59)

5889-2

$2822., f.o.b.

new "DEEP-DIP" RUSTPROOFING

ENGINES:
195.6 CID 6 (90 HP, AMERICAN; 127 HP, SIX; 138 HP - OPTIONAL)
250 CID V8 (215 HP)   (SAME ENGINES IN '59)
327 CID V8 (270 HP, AMBASSADOR)

$1775. ~ 3116. PRICE RANGE

TOTAL 1958 PRODUCTION: 186,227

AMBLER   5902
          5906

$2060. UP STATION WAGON

100 inch wheelbase
bler American

$1835
Suggested delivered price at Kenosha, Wisconsin, for 2-door sedan at
left. State and local taxes, if any,
automatic transmission and optional
equipment, extra.

AMERICAN →

$2145.

5904-1
AMERICAN WAGON
(SUPER) IS new
90 HP

PUSH-BUTTON TRANS. AVAIL.

SUPER →

SIX   5915-1 $2268.

CUSTOM

383.   5915-2

REBEL   CROSS-
108" WB   COUNTRY
          5928-1 or 2
SUPER-1 - $2692.
CUSTOM-2- 2807.

$1821. ~ 3116. PRICE RANGE

59

PRODUCTION:
363,372 (MODEL YEAR)
401,446 (AMC,
CALENDAR YEAR

$2588.

REBEL

5929-2
COUNTRY
CLUB

INTERIOR

$2587.
985-1 or 2 UP

AMBASSADOR

7" WB   270 HP V8

AMBASSADOR
CUSTOM
COUNTRY CLUB
4 DR. H/T →

5989-2

$2822.

DASH

355

RAMBLER

6005 DLX. 2-DR.
$1844.

90 HP — AMERICAN — $2020. 6004

DOORS NOW OPEN WIDER (75° INSTEAD OF 55°)

ROOF RACKS NOW ON ALL WAGONS

6002
$1781. and up

$2098.

6015

60

MODIFIED GRILLE (all except AMERICAN)

6 DELUXE
108" WB
127 HP

6018 WAGON
$2427.

6 SUPER SEDAN
6015-1
6025-1 (V8)
$2268. UP

6015-2 (6)
6025-2 (V8)
2383. UP

CUSTOM COUNTRY CLUB
6019-2 (6)
6029-2 (V8)
$2458. UP

CUSTOM
V8 = 200 HP

new REAR FENDERS

6018 or 6028 (-2 or 4)

REAR COMPARTMENT DETAILS (WAGON)

3 WIDE SEATS, 5 BIG DOORS. The tailgate is a fifth door with outside key lock so children can't open from inside. Rear seat passengers step in—no scrambling over seats or tailgate.

new "COMPOUND WRAP-AROUND" WINDSHIELD on AMBASSADOR

WAGON (6 or 8-PASS.) 6088-1 to 4
2881. and up

$2822. UP

AMB. CUSTOM COUNTRY CLUB 6089-2

AMBASSADOR V-8
BY RAMBLER
The New Standard of Basic Excellence in Luxury Cars

ENGINES:
AMERICAN   195.6 CID 6 (90 HP)
SIX   195.6 CID 6 (127 HP) (OPTIONAL 138 HP)
REBEL   250 CID V8 (200 HP) (" 215 HP)
AMBASSADOR   327 CID V8 (250 HP) (OPT. 270 HP)

434,704 RAMBLERS SOLD 1960.
AMC 1960 CALENDAR YEAR,
TOTAL = 485,745

RAMBLER

6104

3 4-DR. WAGONS also

7-2
-5

$2080.↑

$1894. DLX.

6105

AMER. PRICES
START AT
1831.
(6102)

369.
and up

New! A Convertible

E NEW WORLD STANDARD OF BASIC EXCELLENCE "

new Ceramic-Armored Muffler

$1831. ~ 3111. PRICE RANGE

1961
MODEL YEAR TOTAL: 370,685

AMERICAN
6
L-HEAD
90 (OR
125 HP
OHV IN

AMERICAN
WAGON
REAR
DETAILS

61
(RESTYLED)

SIX, REBEL
REPLACED BY
CLASSIC 6 and V8

new
CLASSIC 6 or V8
127 or 138 HP 6
6

$2098. UP

OR
200 or 215 HP
V8
$2816. UP

CLASSIC FRONT END ↑

CLASSIC
DELUXE

CLASSIC
CUSTOM
6108-2 (6)
6128-4 (V8)
(CLASSIC 2-DR. WAGONS also)

...New! First acoustical ceiling
of molded fiber glass

8 -1,2 or 4

$2841. UP

CUSTOM 400 SEDAN JOINS
AMBASSADOR LINE

AMBASSADOR V8
250 or 270 HP
$2682. UP

6185-5

Rambler
World Standard of Compact Car Excellence

357

RAMBLER

6206

$1846.
90 HP 6

1962 RAMBLER AMERICAN
DELUXE 2-DOOR CLUB SEDAN
(Also offered in Custom and "400" series)

1962 RAMBLER AMERICAN
DELUXE 4-DOOR SEDAN
(Also offered in Custom series)

6205

$1895.

6208 **2130**

RAMBLER AMERICAN DELUXE 4-DOOR STATION WA
(Also offered in Custom series)

$2344.

6208

$2130.
RAMBLER AMERICAN "400" 4-DOOR STATION WAGON
(Also offered in Deluxe and Custom series)

AMERICAN   "400" CVT.

6207-

$1832. ~ 3023. PRICE RANGE

$2150.

6216-2

1962 RAMBLER CLASSIC CUSTOM 2-DOOR CLUB SEDAN
(Also offered in Deluxe and "400" series)

new
GRILLES **62**

ENGINES:
195.6 cid 6 (90, 125, 127 or 138 HP)
327 cid V8 (250 or 270 HP)

new DOUBLE SAFETY BRAKES *with*
TANDEM MASTER CYLINDER

$2349.

6215-5

CLASSIC
400

6218-5

1962
CLASSICS
ARE
6 CYL.
(127 or 138 HP)

$2640.
RAMBLER CLASSIC 6 "400" CROSS COUNTRY STATION WAG

$2760.
6288-2

1962 RAMBLER AMBASSADOR
CUSTOM 4-DOOR STATION WAGON

WB CUT TO 108

250 or 270 HP
AMB. V8s

AMBASSADOR 40
AND INTERIOR

$260

6285-5

AMC PROD., 1962 CALENDAR YEAR: 454,784   1962 RAMBLERS, MODEL YEAR: 423,104

358

RAMBLER

TOP QUALITY AT AMERICA'S LOWEST PRICE!
Manufacturer's suggested retail price for the '63 Rambler American "220" Two-Door Sedan. Optional equipment, transportation, and state and local taxes, if any, extra. An award-winning Rambler value!
$1846

(129,655 AMERICANS BLT.)

6302
6304
220 →
← 220
$1895.
6305

$2281.
6309-7

$2081.

440-H H/T
(with 138-HP OHV 6)
440 CVT.

AMERICAN 6
100" WB

6306-5
440

$2040.
$2245.

6307-5

$2344.

63
(CLASSICS and AMB. TOTALLY RESTYLED)

DASH

6315-2

660

CLASSIC
6 or V8*
new 112" WB
* OPTIONAL 198-HP V8 (STARTING 3-1-63)

770

770

6315-5
2349.
(321,916 CLASSICS BLT.)

New! Hidden storage compartment in wagon!

6318-5 $2640.

SLOGAN:

The New Shape Of Quality

250 or 270 HP
AMBASSADOR V8
new 112" WB

880 CROSS COUNTRY

6388-2

New! Curved glass side windows ... far easier entry!

990 WAGON SIMILAR

$2606.

$2815.

990

6386-5
2-DR.
1963

(28,794 AMBASSADORS BLT.)

AMBASSADOR has LOWER BODY BAND

6385-5
SEDAN
$2660.

WINNER OF MOTOR TREND AWARD
CAR OF THE YEAR

$1832. ~ 3018. PRICE RANGE

EXCEPT FOR '63½ CLASSIC V8,
(SAME ENGINES and HP AS IN 1962)
RAMBLERS, 1963 MODEL YR.: 428,346
AMC TOTAL PRODUCTION,
1963 CALENDAR YEAR: 480,365

359

RAMBLER

new AMER. WAGON RESEMBLES A SEDAN with GRAFTED-ON REAR SECTION

220

6406

220

6408

220
6407-5

AMERICAN

$2057.   330

6405-2

6.00 x 14 TIRES (15" OPT.)

440-H
6409-7

CVT. TOP IN BLK., WHITE, GOLD OR TURQ.

64-440

440
TO 138 HP

$2446.
6418

AMERICAN
TOTALLY RESTYLED
(new 106" WB)

64 440H

$2292.

6418-5

$2651.

770

660 2-DR.

550

CLASSIC (6 has 127 OR 138 HP)
112" WB

64

$2206.

6416-2

770

$2360.

4-DR.
6415-5

6489-5   990-H (INTERIOR BELOW)

6488-5

64-990

990

FRESH NEW SPIRIT OF '64!

DASH (W/O AIR CONDITIONING) →

$2985.

DASH (WITH AIR CONDITIONING) →

990
250 OR 270 HP

112" WB

AMBASSADOR

RAMBLER 1964 PRODUCTION : 379,412
(MODEL YEAR)
AMC CALENDAR YEAR PRODUCTION : 393,863
CLASSIC = 321,916   AMBASSADOR = 28,794
AMERICAN = 151,969

ENGINES :
195.6 CID 6 (90, 125, 127, 138)
232 CID 6 (145 HP IN CLASSIC "TYPHOON" H/T)
287 CID V8 (198 HP)
327 CID V8 (250 OR 270 HP)

RAMBLER

220

$2396.

6508-2

6509-7

American 440-H

06

79.

American 6

AMERICAN

330

2418.
6507-5

ew! 3 different sizes of cars
ew! 3 different wheelbases
ew! 7 spectacular powerplants:
ew Torque Command Sixes-
ost advanced engines! Big V8's

HEAD
5.6 CID
STILL
AIL. IN
ERICAN
0 HP @
00 RPM)
5 HP OHV ALSO)

65

CVTS. NOW IN ALL 3 LINES

DASH (AMERICAN)
412,736 RAMBLERS SOLD 1965

AMERICAN GRILLE NOW VERTICALLY SPLIT INTO 4 HORIZ. SECTIONS

$2727. →

CVT. (3882 BLT.)
195.6 OR 232 CID OHV 6s with 125 OR 155 HP

440

7-5
5
3 BLT.)

6518-5
Rambler Classic 770 Station Wagon

$2436.

SEE ALSO : MARLIN

770

CLASSIC 770

6519-5
Rambler Classic 770 Hardtop

CLASSIC 770 SEDAN
DASH

199 OR 232 CID 6
(128,145,155 HP)
198 HP
with 287 CID
V8

RILLE
LOSE-
JPS

CLAS.

ALSO AVAIL. with 327 CID, 270-HP V8 ALSO USED IN AMB.)

AMB.
(new HEADLIGHTS VERTICALLY STACKED)

6515-5

6587-5

CLASSIC 112" WB CONT'D., BUT AMBASSADOR WB INCREASED TO 116".

6589-7

990

AMBASSADOR V8

$2837. →

SLOGAN :

THE SENSIBLE SPECTACULARS

990-H
DASH

MBASSADOR

1965 RAMBLER MODEL YEAR PRODUCTION: 324,669

AMC CALENDAR YEAR PRODUCTION: 346,367

SEE "AMERICAN MOTORS" SECTION FOR '65 TO '69 MODELS.

$1979. ~ 2955. PRICE RANGE

Rambler American

106" WB (THROUGH '69)

440 WAGON

$2477.

220

(STANDARD MODEL DOES NOT HAVE CHROME STRIP ALONG SIDES.

66

AMC DEALERS ADVERTISED AS THE "FRIENDLY GIANT KILLERS"

440 CVT. $2486.
$2704. (WEST COAST)

RACING STRIPES ON HOOD ARE NOT STD. EQUIPMENT.

$2588. (WEST COAST)

$2370.

GRILLE NOW SPLIT INTO ONLY **3** HORIZONTAL SECTIONS.

199 CID 6 (128 HP) OR ROGUE 290 CID V8 (200 HP)
6.45/6.95 x 14 TIRES

new **Rogue** H/T

6 OR V8

TOTAL 1966 SALES = 265,712

440 4-DR.

$2611. CVT.

$2259.

ROGUE

THIS '67 ROGUE IS THE FINAL CONVERTIBL IN AMERICAN SERIES.

$2073.

DASH

67

220 2-DR.

1967

HORIZONTAL GRILLE PCS. ARE NOW UNBROKEN, (AS THEY WERE IN 1964.)
TOTAL 1967 SALES = 237,785

199 CID 6 (128 HP); 232 CID 6 (145 HP) or 290 CID V8 (225 HP)

DASH

American

4-DR. **2024.**

220

1946.
220
2-DR.
$2179.
(WEST COAST)

68

(INTRO. 9-26-67)

(259,346 SOLD)

new SINGLE HEAVY HORIZONTAL CHROME STRIP ACROSS GRILLE (OTHER HORIZ. PCS. ARE BLACK)

68-220

440 AND **ROGUE** HAVE BRIGHT METAL BETWEEN TAIL LTS.

106" WB CONT'D.

STD. 199 CID 6 (128 HP)
290 CID V8
(200 OR 225 HP)
232 CID 6
STD. IN "ROGUE" (145 HP)
BIG new 390 CID V8 (315 HP) IN "SC/ RAMBLER HURST"

(FINAL USE OF "RAMBLER AMERICAN" NAME)
REPLACED 9-25-69 BY AMC "HORNET" FOR 1970

69

(INTRO. 10-1-68)

"RAMBLER" NAME NO LONGER APPEARS ON GRILLE. SOME VERTICAL PCS. NOW ALSO VISIBLE.

Rambler $1,998
(SALE PRICE)
(RAMBLER 220, 440 and ROGUE RANGE IS $2231. TO $2710.

new TRI-COLOR HOOD EMBLEM

STUDEBAKER CORP., SOUTH BEND, IN.
169.6 cid 6 (80 HP)

STUDEBAKER

(1902~1966)

"DOUBLE DATER" COUPE

DASH

SKYWAY CHAMPION is ONLY SERIES AVAIL.

EARLY **46** "5-G"

RARE!

(AVAIL. ONLY TO MAY, '46)

110" WB

$ **916**.. f.o.b.

(1285 BLT.)

BODIES AVAIL.= 3-PASS. COUPE; (2465 BLT.)
5-PASS. "DOUBLE DATER" CPE.;
2-DR. CLUB SED.; 4-DR. CR. SEDAN
(5000 BLT.)   (10,525 BLT.)

ALL-NEW 1947 MODELS START MAY, 1946

new 112" WB on CHAMPION
6-G → DELUXE SEDAN

169.6 cid 6 (80 HP)

$1478.

47

(TOTALLY RESTYLED)

PRODUCTION
CHAMPION (105,097 BLT.)
COMMANDER (56,399 BLT.)

Starlight COUPE

$1752.

CHAMPION STRLT. CPE. has 1-PC. WINDSHIELD, UNLIKE OTHER CHAMPION MODELS
(SAME HP FIGS. SINCE '41)

14-A
← COMMANDER   SEDAN →

119" WB
226.2 cid 6 (94 HP)

$1761.

COMMANDER REGAL DE LUXE

DETAILS OF 2-DOOR SEDAN (CHAMP. REGAL DE LUXE)

$1910.

REGAL DE LUXE LAND CRUISER

* L.C. PART OF COMMANDER SERIES

123" WB

CONVERTIBLE (new)

CHAMPION   3-W. CPE.

$1535. ↗

new HORIZONTAL PIECE ACROSS EITHER END OF CHAMPION GRILLE

New 1948 Studebaker
First in style

$2060.

(COMMANDER CVT. ALSO AVAIL. AT $2431.)

SEDANS

CHAMPION

$2077.

48 COMMANDER

new HORIZ. CHROME ABOVE CMNDR. GRILLE

$1636.

PRODUCTION 99,282 CHAMPIONS; 85,711 COMMANDERS

"First in style...first in vision...first by far with a postwar car"

364

STUDEBAKER

ENGINES : 169.6 CID 6 (80 HP) OR new 245.6 CID 8 (100 HP)

$1762

CHAMPION (8-G)
NOW has 2 HORIZ. STRIPS ACROSS GRILLE

$1757.

$2135., f.o.b.

(85,604 CHAMPIONS BLT.)

VERTICAL CHROME CENTER STRIP ADDED TO CMNDR. GRILLE

STARLIGHT CPE. and INTERIOR

STUDE. "STARLIGHT" CLUB COUPES ARE AMONG THE MOST UNUSUAL and ATTRACTIVE BODY STYLES EVER PRODUCED!

49

LAND CRUISER INTERIOR

LAND CRUISER

(43,694 CMNDR. and L.C. BLT.)

REAR

CHAMPION (270, 604 BLT.)

lowest price

CHAMPION CUSTOM 6-PASS. 2-DOOR SEDAN AS SHOWN

$1487.50

new 113" WB (CHAMPIONS)

85 HP

50

new "BULLET-NOSE" FRONT END STYLING

"the "next look" in cars "

new 6.40 x 15 TIRES

(CONT'D.)

STUDEBAKER

$1676.

CHAMPION REGAL DE LUXE 6

CHROME ALONG ROCKER PANEL

Studebaker

50
(CONT'D.)

2-DR.

$1566.

CHAMPION DLX. (9-G) has RUBBER PAD ON REAR FENDER, BUT NO CHROME ALONG ROCKER PANEL.

CVT.

America likes Studebaker's new driving thrill—Every 1950 Studebaker handles with light-touch ease—rides so smoothly it almost completely abolishes travel fatigue. A new kind of coil spring front suspension.

COMMANDER (17-A) 120" WB

$2024.

$2013.

(72,562 COMMANDERS BLT., INCLUDING LAND CRUISER)

LAND CRUISER 124" WB

America likes this "next look" in interiors—Fabulously fine nylon cord upholstery, introduced into motoring by Studebaker, is standard in the 1950 Land Cruiser and regal de luxe Commander. Land Cruiser is shown.

$2187.

1950 PRICE RANGE $1419. ~ 2328.
(CHAMP. CUSTOM 3-PASS. CPE.)   (CMNDR. REGAL DLX. CVT.)

STUDEBAKER

new 115" WB (ON ALL BUT LAND CRUISER)

CHAMPION REGAL SEDAN

FIRST YEAR FOR STUDEBAKER V-8
(STD. EQUIP. IN CMNDR., L.C.)

$1833.

CHAMP.
CUSTOM
HAS NO
HOOD
ORNAMENT.
CHAMP.
DELUXE HAS
ORNAMENT.

DLX.
3-WINDOW
BUSINESS
COUPE
$1643. →

L-HEAD
169.6 cid 6
(85 HP)
OR
232.6
and
new O.H.V.
V8 ENGINE
ALSO AVAIL.
(120 HP)

51

Studebaker Champion

6.40 x 15
TIRES

CHAMPION
DE LUXE 6 (10-G)
85 HP

$1744.

(REGAL CHAMP.
has LEATHER TRIM
INSIDE DOORS.)

A brand new V-8
(233 CID)
Commander
STATE CMNDR.
CVT.

$2381. has

120 h.p.

@4000 (THROUGH '54)

STATE
CMNDR. SEDAN

$2143.

COMMANDER
LAND CRUISER
$2289.

"BULLET NOSE"
GRILLE SOMEWHAT
MODIFIED FROM
'50.

new
GRILLE IS
FLUSH
WITH
FRONT
END

FINER PCS. IN
GRILLE

new SHORTER 119" WB ON L.C.

"STUDEBAKER...THE THRIFTY ONE FOR '51"

$1561. ~ 2381.
PRICE RANGE

CHAMPION = 144,286 BLT.; CMNDR./L.C. = 124,280 BLT.
6   V8

367

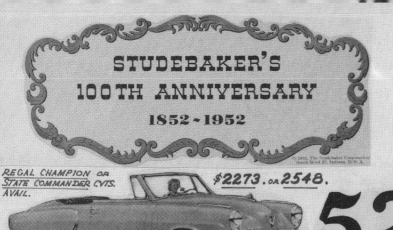

STUDEBAKER'S
100TH ANNIVERSARY
1852~1952

1952, The Studebaker Corporation
South Bend 27, Indiana, U. S. A.

REGAL COMMANDER
$2121.

REGAL CHAMPION OR
STATE COMMANDER CVTS.
AVAIL.

$2273. OR 2548.

PACE CAR
AT 1952 INDY 500 RACE

52
FRONT END
RESTYLED

$2365.

LAND CRUISER

1952 PRODUCTION :
CHAMPION = 101,390
COMMANDER, L.C.= 84,849

CHAMPION
PRICES START AT
$1735.

BODY DESIGN
BASICALLY
AS BEFORE,
BUT
CONTROVERSIAL
"BULLET NOSE"
DISCONTINUED
IN FAVOR OF
A MORE
CONVENTIONAL
(BUT EXTREMELY
BROAD)
GRILLE

"STARLINER"
H/T is new

AVAIL. in CHAMP. OR CMNDR.
SERIES

$2488.

STATE
COMMANDER
STARLINER

See Studebaker for '52

$2172.

(2-DR.)

EDITED PHOTOS (IN ADS)
CAUSE PASSENGERS IN CARS TO
APPEAR UNUSUALLY SMALL, IN
COMPARISON TO THE CAR.
(THIS MAKES THE CAR LOOK
LARGER.)

$2208.

STATE COMMANDER

$1735.~2548. PRICE RANGE

See and drive the Studebaker Starliner—It's America's smartest
"hard-top"—available either as a Champion or a Commander V-8.

368

Studebaker

REGAL CHAMPION STARLINER H/T

CHAMPION PRICES START AT $1735.

85 HP @ 4000 RPM

170 CID 6 (THROUGH '54)

53 $2116.

MEANS CHAMP. 6 H/T

(TOTALLY RESTYLED)

new DESIGN WINS FASHION ACADEMY AWARD

new 116½" WB 120½" on COUPE., H/T and LAND CRUISER)

MEANS CMNDR. V8

$2488.

$1735. ~ 2374. PRICE RANGE

NO MORE CONVERTIBLES AVAILABLE (UNTIL '60 LARK)

1953 PROD.:
CHAMP. = 93,807
COM., L.C. = 76,092

CALENDAR YEAR PRODUCTION = 188,844

6.40 × 15 TIRES (6)
6.70 × 15 WAGON (6)
7.10 × 15 (V8)

1954 PRODUCTION: CHAMPION 51,431
COMMANDER, L.C. 76,092

SAME ENGINES SINCE '51

STATION WAGON INTRODUCED!

$1758. ~ 2556. PRICE RANGE

VERTICAL PIECES ADDED TO GRILLE FOR 1954.

STUDEBAKERS, SINCE LATE 1930s, ARE Styled by Raymond Loewy

CHAMPION

CHAMPION PRICES START AT $1758. (COUPE)

54 new GRILLE

COMMANDER

$2136. UP

new CONESTOGA WAGON (116½" WB) $2556.

LAND CRUISER

$2438.

STUDEBAKER

$1741. ~ 3253. PRICE RANGE

6 CYL. has 101 HP @ 4000 RPM (new 186 CID THROUGH '58)

CHAMPION CUSTOM

A BIG NEW CHAMPION
America's No. 1 economy car!
Now more marvelous than ever!
$2125. REGAL H/T

CHAMPION PRICES START AT $ **1741.**
2-DR.

new GRILLE (with HEAVY CHROME BORDERS)

55

CHAMPION DE LUXE

COUPES

$1875.

INTERIOR (CMNDR.)

$2127.

Now in the low price field!
A sensationally high-powered
NEW COMMANDER V-8

COMMANDER

$2094.

(6-CYL. OR V8 WAGONS)

COMMANDER REGAL
CONESTOGA

SPEEDSTER

$3253.

$2445.

COMMANDER H/T DETAILS

$2282.

STATE PRESIDENT
$2456.

NEW!

AMERICA'S SMARTEST TWO-TONING!

PRESIDENT

The first dynamic headliners of the great Studebaker-Packard alliance! Sensationally powered '55 Studebakers! Amazingly low introductory prices!

$2381.

ENGINES: new 185.6 CID 6 (101 HP) CHAMPION
new 224.3 CID V8 (140 HP) EARLY '55 COMMANDER
new 259.2 CID V8 (162 HP) LATER " " (175 HP) PRESIDENT
(185 HP) EARLY '55 SPEEDSTER; " " PRESIDENT

1955 PRODUCTION:
CHAMPION 6 = 50,368
COMMANDER V8 = 58,792
NEW PRESIDENT V8 = 24,666

370

CHAMPION PRICES START AT $1841.

Studebaker

Craftsmanship with a Flair

PELHAM 6

CHAMPION 6 $1885.

wagons

$2232.

PARKVIEW V8 $2354.

FLIGHT HAWK 6

CMNDR. V8 $1986.

$2101.

POWER HAWK V8

$3610.

SKY HAWK V8

FRONT DETAILS OF COMMANDER V8

NEW DUAL EXHAUSTS · Built into the bumper for more style, more class than you've ever seen in a low price car. Ready for 4-barrel carburetion to boost mileage and power.

Hawks $4071.

(3610 BLT.)

GOLDEN HAWK V8 (275 HP)

The Golden Hawk

FRONTS RESTYLED **56**

$2235. SEDAN

PRESIDENT V8

CLASSIC SEDAN *has* ROCKER PANEL TRIM

(4071 BLT.)

"CLASSIC" SEDAN $2489.

1958

259, 289 OR 352 CID V8s

PRESIDENT V8 195 HP * @ 4500 RPM

new 12-VOLT ELECTRICAL SYSTEM

NEW CYCLOPS-EYE SPEEDOMETER

* 210 HP *with* 4-BBL. CARB.

Prosident

PRESIDENT PINEHURST V8

$2529.

ENGINES:
185.6 CID 6 (101 HP)
259.2 V8 (170 OR 185 HP)
289 CID V8 (195, 210 OR 225 HP)
352 CID V8 (275 HP)

CALENDAR YEAR PRODUCTION: 82,955

$1741. ~ 4071. PRICE RANGE

Studebaker

CHAMP. SCOTSMAN IS A *new* BUDGET-PRICED MODEL with MINIMUM of CHROME and PLAINEST INTERIOR

101 HP

W-1 SEDAN

2-DR., 6-CYL. WAGONS
SCOTSMAN (116½" WB)
PELHAM (118½" WB)

57-G
CHAMPION SCOTSMAN 6 (new)
PRICES START AT $1776. (2-DR.)

PLAIN, PAINTED HUB CAPS, 6.40 × 15 TIRES

CHAMPION DE LUXE 6 $2171.

57 (RESTYLED)

116½" WB ON MOST

57-G (6 CYL.)
57-H (8 CYL.)

$2505.

PROVINCIAL 4 DR. WAGON P-4

D-4
PARKVIEW 2-DR. WAGON
COMMANDER
F-2 (CUSTOM)
F-4 (DELUXE)

$2561.

$2407.
PRESIDENT W-6

$2246. (DLX.)

Studebaker-Packard

CORPORATION

Where pride of Workmanship comes first!

P-6
BROADMOOR 4-DR. WAGON
$2666.

170 HP V8 RAISED TO 180 HP
185 HP V8 RAISED TO 195 HP

OTHER ENGINES AS IN 1956

SILVER HAWK (15,318 BLT.)
GOLDEN HAWK (4356 ")
TOTAL 1957 STUDEBAKER PRODUCTION:
74,738   (CALENDAR YEAR: 67,394)
(1957 HAWKS on NEXT PAGE)

STUDEBAKER

$2263.

C-3
SILVER HAWK COUPE
186 CID 6
(101 HP)

289 CID (275 HP @ 4800 RPM)
V8
GOLDEN
HAWK
H/T

K-7

$3185.

OR
210, 225 HP V8s
(289 CID)

$57
(CONT'D.)

Golden Hawk

(10,325 CHAMPIONS BLT.)

Studebaker Commanders
and Champions

58
4 HEADLIGHTS
ON SOME
MODELS

101 TO
275 HP

(SAME HP AS '57)
THE FINAL GOLDEN HAWK (878 BLT.)
$3282.

180-HP CMNDR.

" Studebaker cars take on a
completely new luxury look for 1958!"

SCOTSMAN 6
PRICES START AT
(20,870
SCOTSMANS
BLT.)
NOTICE
STRIKING
DIFFERENCES
IN APPEARANCE
BETWEEN THE
LT.-OVER-DK.
AND
DK.-OVER-LT.
HARDTOPS

$1795.

REAR DETAILS
(SEDAN)

$2639.
Studebaker
President

The Hawk-inspired PRESIDENT
STARLIGHT for 1958
H/T

$2695.

Studebaker-Packard
CORPORATION
Where pride of Workmanship comes first!

10,442 PRESIDENTS BLT. (4 DR. SEDAN and H/T)

CALENDAR YEAR PRODUCTION:
55,175

373

STUDEBAKER

59

HAWK 6 PRICES START AT **$2360.**

6-(2417 BLT.)
V8-(5371 ")

170 CID 6 (90 HP @ 4000) OR 259 CID V8s (180 OR 195 HP @ 4500)

6 OR V8

SILVER HAWK (C-6)

1959

SEE ALSO: **LARK** (STARTING 1959, LOWER-PRICED MODELS USE LARK NAME)

$2495. (V8)

120 1/2" WB

(3939 BLT.)

3 new STRIPS on SIDE of REAR FENDER

60 new 289 CID V8 RETURNS TO HAWK AS ONLY AVAIL. ENG. (210 OR 225 HP @ 4500 RPM)

$2650.

C-6 HAWK

6.70 x 15 TIRES

6.70 x 15" TIRES

61 new TRIM DESIGN ALONG REAR FENDERS

$2650.
$2677.

(3340 BLT.)

HAWK

(C-6)

new HAWK GT

new ROOFLINE 120 1/2" WB

(8388 BLT.)

62 (RESTYLED)

new CLASSIC-STYLE GRILLE with HEAVY CHROME BORDERS

$3424. (UP $27. IN '63)
WEST COAST

(K-6) $3095.

new GRILLE DESIGN with DECORATIVE CRISS-CROSS STRIPS ADDED

$3095.

HAWK GT

63

$4445. new AVANTI

(4634 BLT.)

WEST COAST $4759.

AVANTI INTRO. DURING '62   109" WB
289 CID V8 (240 OR 290 HP)

(3834 BLT.)

STUDEBAKER (and LARK) **CALENDAR YEAR PRODUCTION:**

1959 = 153,823	1960 = 105,902	1961 = 78,664
1962 = 86,974	1963 = 67,918	

HAWK GT (1963) HAS 289 CID V8 (210, 225, 240 OR 290 HP)

STUDEBAKER

64

$4445.

ENGINES:
169.6 cid 6
(112 HP)
259.2 cid V8
(180 or 195 HP)
289 cid V8 (210,
225, 240, 290 HP)
new 304½ cid V8
(280 or 335 HP)

109" WB

AVANTI V8

(809 BLT.)

6.00, 6.50 or 6.70 × 15 TIRES

"LARK" DESIGNATION PHASED OUT DURING 1964.

113" WB

109" WB

COMMANDER
$2303.

CHALLENGER (109" WB)

(AVAIL. 1964 ONLY)
112 HP 6 or
180 HP V8

$2048. UP
$2417. WEST COAST

113" WB on
CRUISER V8

120" WB

1964

FINAL
GRAN TURISMO HAWK
(1767 BLT.)

$2966.

TOTAL 1965 PRODUCTION: 19,435
$2125. ~ 2890. PRICE RANGE

CHEVROLET ENGINES NOW USED
194 cid 6 (120 HP)
283 cid V8 (195 HP)
(SAME ENGINES IN 1966)

STUDE. PRODUCTION CONTINUES ONLY AT THE CANADIAN BRANCH FACTORY, FOR '65-66, (HAMILTON, ONT.)

DETAILS OF WAGONAIRE ILLUSTR. AT RIGHT

BRAKE RELEASE

TOPSIDE LUGGAGE RACK (OPT.)

Wagonaire
(V8) $2695.

65

6 CYL. CHEV. OHV ENG.
(V8 on NEXT PG.)

COMMANDER

"the Common-Sense Car"

TAILGATE STEP

(UNFOLDS AND LOWERS)

$2230.
$2581. (6 CYL.)
WEST COAST

(CONT'D.)

INTERIOR

Studebaker
THE COMMON-SENSE CAR

Exclusive Beauty Vanity in glove compartment—(opt.).

Daytona Sports SEDAN

65 (CONT'D.)

$2405. (6)
2500. (V8)

DAYTONA INTER.

CLOSER DETAIL OF DASH

V8 NOW has 195 HP

← FINAL YEAR of 4 HEADLIGHTS
$2610. →
WEST COAST:
$2985. →

Cruiser

FINAL STUDEBAKERS HAVE THIS GRILLE.

S = 6
V = V8

WEST COAST
PRICES START AT $2060. $2465. (COMMANDER 6 2-DR.)

194 CID, 120-HP 6
or 283 CID, 195-HP V8

LAST

Studebaker
AUTOMOTIVE SALES CORPORATION

Cars BY ST.-P.

66

$2555. UP

109" OR 113" WB (SINCE 1962, ON LARK and LARK-BASED MODELS)

TOTAL 1966 PRODUCTION = 8947
$2060.~2695. PRICE RANGE

STUDEBAKERS DISCONTINUED MARCH, 1966

TEMPEST BY PONTIAC!

new COMPACT CAR

new SLANTED 4-CYL. ENG. ADAPTED FROM THE RIGHT HALF OF A PONTIAC V8!

FROM $2329. WEST COAST

new FOR **61** *new!*

61½ COUPE ROOF LINE

COUPES (STD. 6 PASS. or DLX. BUCKET SEATS) INTRO. IN MIDYEAR

112" wheelbase (THROUGH '63)

THE HOT TOPIC IS THE NEW TEMPEST BY PONTIAC

TROPHY 4 ENGINE

FOUR CYLINDERS

to 155 h.p. (Or buy the 155 h.p. aluminum V-8 option.)

110 or 130 HP (STD.)

4 (194.5 CID)

OR

V-8 (215 CID) 155 HP

COUPE $2113. UP

Independent suspension at all wheels

CUSTOM (DLX.) '61½ COUPE has "TOWN CAR" REAR WINDOW.

FRONT ENGINE ⟺ REAR TRANSMISSION

PERFECT BALANCE

PONTIAC'S TEMPEST
PICKED BY MOTOR TREND MAGAZINE AS
CAR OF THE YEAR

WITH
PONTIAC POWER STEERING (OPT.)

SAFARI WAGON (REAR 3/4 VIEW) $2438. TO 2622.

$2113. ~ 2622. PRICE RANGE

CALENDAR YEAR PRODUCTION = 115,945

TEMPEST

STANDARD TEMPEST COUPE has BROAD BACKLIGHT, MINIMUM CHROME

$2186.

$2240.

new GRILLE

62

CUSTOM COUPE has "TOWN CAR" BACKLIGHT

$2297.

194½ CID 4 CYL. WITH 110, 115, 120, 140, OR 166 HP. 185-HP ALUMINUM V8 ALSO AVAIL. (215 CID)

new LE MANS $3200. (WEST COAST)

STD. WAGON

$2511. UP

The gas-saving "4" with Pontiac Punch!

LE MANS

MODELS :		TOTAL 1962
COUPE (15,473 STD.)	4 DR. WAGON (6504 STD.; 11,170 DLX.)	PRODUCTION:
SPT. ", CUSTOM (12,319 DLX.; 39,662 LE MANS)	CONVERTIBLE (5076 DLX.; 15,559 LE MANS)	143,193
4-DR. (16,057 STD.; 21,373 DLX.)	(CUSTOM = DLX.)	

new TAIL LIGHTS

4 CYL. 195.4 CID (115-166 HP) TEMPEST (NAME RETURNS TO FRONT FENDER)

(10,135 BLT.)

$2512. UP

$2241. UP

AR-9186

(28,221 BLT.)

CVT. WITH TOP DOWN $2742.

LE MANS

new 326 CID V8 ALSO AVAILABLE (260 HP)

$2418. (45,701 BLT.)

LE MANS has RECTANGULAR TAIL-LIGHTS, "LE MANS" ON FRONT FENDER.

new GRILLE

63

LE MANS CVT. (15,957 BLT.)

WITH TOP UP

TOTAL 1963 PRODUCTION: 143,6/6

Wide-Track Pontiac Tempest

TEMPEST $2313. SEDAN (19,427 BLT.)

$2259. S.PT. COUPE (6365 BLT.)

64 (RESTYLED)

Tempest

new 115" WB
new 215 CID
140-HP IN-LINE
O.H.V. 6

Tempest CUSTOM $2345.

SAFARI WAGON $2605. UP

CUST. SPT. CPE. (25,833 BLT.)

326 CID V8 ALSO AVAIL. (250 OR 280 HP)

LE MANS SPT. CPE. $2491.

(29,948 BLT.)
(31,317 BLT.)

$2399.

(H/T $2556.)

$2641.

CUSTOM CONVERTIBLE (7987 BLT.)

1964

GTO ("GTO" APPEARS ON GRILLE) (new)

LE MANS

CVT. (17,559 BLT.) $2796.

REAR DETAILS

1964

$2259. ~ 3500. PRICE RANGE
TOTAL 1964 PRODUCTION: 250,328

GTO SPT. CPE. (7384 BLT.) – $3200.
H/T (18,422 BLT.) ——— 3250.
CONVT. (8644 BLT.) ——— 3500.

TEMPEST

1965: The year of the Quick Wide-Tracks

$2313.
(15,705 BLT.)

$2605.

223
TEMPEST

SAFARI WAGON
(5622 BLT.)

Tempest

FROM
$2618.
(WEST COAST)

65

new GRILLES
new TAIL LIGHTS

H/T
(new)

CUSTOM H/T
(29,906 BLT.)
$2411.

PONTIAC TEMPEST

140-HP 6 (215 CID)
OR
250-285 HP V8
(326 CID)

235
TEMPEST CUSTOM

4-DR.
(25,242 BLT.)
$2400.

SAFARI
4-DR. WAGON
(10,792 BLT.)
$2619.

LE MANS H/T
(60,548 BLT.)

$2556.

LE MANS SPT. CPE.
(18,881 BLT.) $2491.

237
Le Mans

CVT.
(13,897 BLT.)
$2797.

FRONT-END COMPARISON
OF LE MANS (left) and GTO (right)

GTO CVT.
(11,311 BLT.)

SEE
ALSO:
Pontiac

OFFICIAL PACE CAR-MOTOR TREND RIVERSIDE "500"
COURTESY OF
HURST

$3057.

GTO

TOTAL 1965 PRODUCTION :
326,019

GTO H/T (55,722 BLT.)
(GTO SPT. CPE. AVAIL.- 8319 BLT.-$2751.)

$2816.

$2260. ~ 3057. PRICE RANGE

TEMPEST OHC SPRINT

← PLAIN TAIL LTS. ON LOW PRICED MODELS

1966 PRODUCTION: 384,794

$2278.~3082. PRICE RANGE

(230 CID OHC 6, 165 HP)
207 HP @ 5200 RPM, SPRINT
OR 326 CID V8 (250 HP)

(160-MPH SPEEDOMETER and TACH. on SPRINT)

115" WB

TEMPEST

$2568. LE MANS

$3006. (H/T)

TEMPEST (SPRINT PKG.)

CVT- FROM $3093. (WEST) $2665. (BASE PRICE)

(5557 BLT.)

66

(TEMPEST, CUSTOM, LE MANS OR GTO MODELS)

(REAR) GTO

(78,109 BLT.)

GTO ↗ ↓

GTO has RALLY LTS., "GTO" on GRILLE.

GTO has 389 CID V8 (335 HP)

"GTO" MEANS "GRAN TURISMO OMOLOGATO"

CVT. $3425. (WEST)

(12,798 BLT.)

The tiger scores again! Wide-Track Pontiac/'66

1967 PRODUCTION: 288,924

$2341. ~ 3165. PRICE RANGE

$3165.

Wide-Track Pontiac/67

GTO CVT. (9517 BLT.) $3547. (WEST)

HOOD SCOOP DETAIL

7.75 x 14 TIRES

LE MANS

SAME HP AS 1966, EXCEPT new 215 HP IN SPRINT

$2935.

H/T $3317. (WEST)

Pontiac GTO

67

has F 70 x 14 TIRES

(65,176 BLT.)

RALLY I

new GRILLES (TEMPEST FROM $2787.)

RALLY II

H/T $3094. (WEST)

$2648. (75 965 BLT.)

381

250 cid OHC 6 (175 or 215 HP)
350 cid V8 (265 or 320 HP)
or 400 cid V8
(265, 350 or 360 HP)

TEMPEST

LE MANS

7.75 / 8.25 x 14" TIRES

H/T
(110,036 BLT.)

$2786.

new CONCEALED
WINDSHIELD
WIPERS

TEMPEST
and
TEMPEST
CUSTOM
SHARE
LE MANS
STYLING.
$2509. UP

2 DR. =
112" WB
116" = 4 DR.

$3101.

RESTYLED **68**

GTO has 400 cid V8 (4 BBL)
(265 or 350 HP)
3-SPEED HURST SHIFTER
AVAIL.

H/T
(77,704 BLT.)

GMAC
TIME PAYMENT
PLAN
General Motors Acceptance Corporation

↙ GTO
↓

GTO
has
new
CONCEALED
HEAD-
LIGHTS
CVT.
(9980 BLT.)

$3227.

We've just received our 4th Car of the Year award.

Wide-Track 1968 Pontiacs

382

TEMPEST

400 CID V8 (366-370 HP)
ALL RISE FOR

THE JUDGE

"ENDURA" SCRATCH/DENT-RESISTANT
FRONT END IS BODY-COLORED

SPOILER

The Judge: a special GTO by Pontiac

GTO 3/56.

H/T $3544. (WEST)

69
new GRILLES

250 CID 6 (175 HP)
STD. 350 CID V8 (265 HP)

DASH (JUDGE)
RED-ORANGE IS STD. COLOR OF "JUDGE"

H/T (82,817 BLT.)
LE MANS
$2835.

H/T $3292. (WEST)

The year of the great Pontiac break away

$2510. ~ 3382. PRICE RANGE

WATER SPORTS

ON-HOOD TACHOMETER *and* HURST SHIFTER ILLUSTRATED

GTO
CONVERTIBLE (7436 BLT.)

G78 x 14" TIRES
400 CID V8 (350 HP)
STD. IN GTO *

(* = 366 OR 370 HP OPT.)

$3382.
$3770. (WEST)

383

THUNDERBIRD

(Ford)

(INTRO. FALL, 1954, FOR 1955)

(MODEL 40)

ALL *with* V-8 O.H.V. ENGINES (292 cid)

55

"CLASSIC" T-BIRDS AVAIL. *with* REMOVABLE HARD TOP OR CVT. TOP (THROUGH '57)

193 HP (198 HP w. A.T.)

16,155 BLT.

102" WB (THROUGH '57)

6.70 × 15 TIRES (THROUGH '56)

$2944.

15,631 BLT.

$3151.

2 ENGINES NOW AVAILABLE: 292 cid V8 (202 HP) or new 312 cid V8 (215 HP, overdrive) (225 HP, A/T)

INTERIOR

40-B

REMOVABLE H/T *has* new PORTHOLES (EXCEPT EARLY MODELS)

56

40-A

new 7.50 × 14 TIRES

57

NAME MOVED TO FRONT FENDERS

new GRILLE; BUMPERS, TAIL-LIGHTS MODIFIED

$3408.

new WHEEL COVERS

new DASH

21,380 BLT.

292 cid V8 (212 HP) or 312 cid V8 (245, 270, 285 or 300 HP)

THUNDERBIRD

new H/T (35,758 BLT.) $3631.

NOW with 4 HEADLIGHTS

58 (TOTALLY RESTYLED)

DASH

63-A

Exclusive "Panel Console"   76-A

new 113" WB $3929.

CVT. (2,134 BLT.)

REAR DETAILS

1958-T

300 HP (THROUGH '65)
new 352 CID V8

The car everyone *would love to own!*

H/T 63-A

(57,195 BLT.)

CVT. 76-A

$3696.

$3979.

new 8.00×14 TIRES

59

HORIZONTAL PCS. ON new GRILLE and BETWEEN TAIL-LIGHTS

new 430 CID V8 (350 HP) OPTIONAL

T-59

new SIDE TRIM EACH YEAR

(VT. 10,261 BLT.)

6 TAIL-LIGHTS IN 1960

9 VERTICAL CHROME BANDS on EA. REAR FENDER (1960 ONLY)

$4222. CVT.

60

1960

REAR DETAILS

new GRILLE DETAILS

new GRILLE, 6 ROUND TAIL LIGHTS

sliding sun roof (new)

$3755.

'60 THUNDERBIRD
THE WORLD'S MOST WANTED CAR

63A H/T (78,447 BLT.)
(ALSO AVAIL. = 63-B GOLD TOP H/T = 2536 BLT.)
APPROX. $3900.

76A CONVERTIBLE (11,860 BLT.)

TOTAL 1960 PRODUCTION: 92,843

THUNDERBIRD

63A H/T (62,535 BLT.)
76A CVT. (10,516 BLT.)

ALL HAVE 390 cid V8 (300 HP)

'61 THUNDERBIRD
UNIQUE IN ALL THE WORLD

CVT. $ **4637.**

PACE CAR AT 1961 INDY 500 RACE

← (OPTIONAL)

Swing-Away Steering Wheel glides out of your way for easier, more graceful entrances and exits—yet locks safely in place before you can drive.

61
(TOTALLY RESTYLED)

H/T $ **4170.**

8.00 x 14" TIRES

unmistakably New, unmistakably Thunderbird

$4321.

new BODY TYPES and MODEL NUMBERS in 1962

$4398.

LANDAU (new)

HARDTOP

$**5552.**

CVT. (7030 BLT.)

ORIG. $5439.

with VINYL TOP and DECORATIVE LANDAU IRONS

$4511.
(WEST COAST)

Thunderbird
Sports Roadster
(new)
(1427 BLT.)

new GRILLE

62

new SPTS. RDST. has TWIN TONNEAU CAPS (as illustrated)

(69 554 BLT.)
H/Ts = MODEL 83
CVTS. = MODEL 85
(8457 BLT.)

unique in all the world

390 cid V8 (300 or 340 HP)

(SAME ENGINES AS IN 1962)

THUNDERBIRD

$4529. TO $5648.
WEST COAST PRICE RANGE

H/T 83   (42,806 BLT.)

(14,139 BLT.)

$4445.

113" WB

63

INTERIOR (OFFERING WOOD-GRAIN EFFECTS)

LANDAU $4548.
87

CVT. 85

(5913 BLT.) $4912.

final SPORT ROADSTER 89
(455 BLT.)

$4445.~5563. PRICE RANGE

390 CID V8 (300 HP) (THROUGH '65)

113.2" WB (THROUGH '66)

H/T $4486.
(60,552 BLT.)

64
(RESTYLED)

CVT. (9198 BLT.)
$4953.
new 8.15 × 15 TIRES

new WIDE TAIL-LIGHTS with T-BIRD EMBLEM $4589.

LANDAU H/T (22,715 BLT.)

new DASH

PRICED FROM $4486.

(IN '64 and '65)

H/T (42,652 BLT.)

LIGHT OR DARK TOP COVERING AVAIL.

LANDAU

INTERIOR with new WOOD-GRAIN EFFECTS

CONVERTIBLE (6846 BLT.)
$4953.

new 5 VERTICAL STRIPS on EACH TAIL-LIGHT

65

2 VERSIONS OF LANDAU H/Ts :
LANDAU (STD.) $4589.
(20,974 BLT.)
LIMITED ED. SPEC. LANDAU
(4500 BLT.) $4639.

GRILLE SLIGHTLY CHANGED.
THUNDERBIRD EMBLEM AT FRONT and BACK END, INSTEAD OF NAME.

FINAL THUNDERBIRD CONVERTIBLE PRICED AT **$4879.** (5049 BLT.)

$4483.

THUNDERBIRD

113" WB

(15,633 BLT.)

$5005. (WEST COAST)

TOWN HARDTOP

66 new GRILLE; new TAIL-LIGHTS

Highway Pilot

CONTROL

TOWN LANDAU H/T (35,105 BLT.)

H/T (13,389 BLT.) $4426.

$4584. ↑

SEQUENTIAL TURN SIGNALS IN FULL-WIDTH TAIL-LIGHTS

PROD.: 69,176 (CALENDAR YR., 72,734)

390 C/D V8 (315 HP)

TOWN LANDAU H/T TOP DETAILS

Stereo-Tape System... Overhead Safety Control Panel

(IN TOWN H/T) OVERHEAD CONSOLE WITH WARNING LIGHTS →

1967 PROD.: 77,956 (CALENDAR YR., 59,640)

H/T (15,567 BLT.)

(WEST COAST FR. $5144.)

4 DR. (new)

$4603.

(24,967 BLT.)

note HOW A SIDE SECTION OF TOP OPENS WITH REAR DOOR

DASH

67 (RESTYLED) (INTRO. 9-30-66)

(4 DR. DETAILS on NEXT PAGE)

388

THUNDERBIRD

$4825.

67 (CONT'D.)
(LANDAU 2-DR. ALSO AVAIL.)

new CONCEALED HEADLIGHTS IN *new* GRILLE

new 115" WB (2-DR.)
11.7" WB (4-DR.)

(WEST COAST) $5366.

LANDAU 4-DOOR IS *new*

INTERIOR   1968 PRODUCTION: 64,931

(INTRO. 9-22-67)

68

H/T 2-DR. FROM $5263. (WEST)

new SIDE SAFETY LIGHTS

$4924.

new GRILLE

new 429 CID V8 (360 HP)

8.15/8.45 × 15 TIRES

4-DR. LANDAU $5471. (WEST)

$4964.

OPTIONAL SUN ROOF

2-DR. $5359.
2-DR. LANDAU 5499. } (WESTERN PRICES)
4-DR. LANDAU 5578.

1969 PRODUCTION: 49,272

2-DR. (FORMAL ROOFLINE) LANDAU

(27,664 BLT.)

new 8.55 × 15 TIRES

69
new GRILLE

(INTRO. 9-27-68)

INTERIOR

STD. 2-DR. ROOFLINE (5913 BLT.)   4-DR. (15,695 BLT.)

389

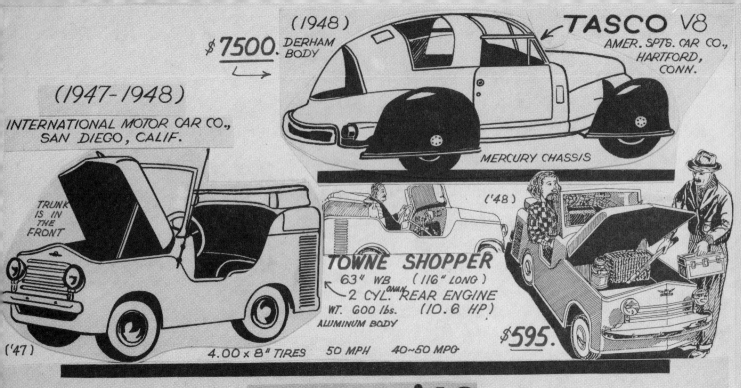

TASCO V8
AMER. SPTS. CAR CO.,
HARTFORD,
CONN.

(1948)
DERHAM BODY

$7500.

MERCURY CHASSIS

(1947-1948)
INTERNATIONAL MOTOR CAR CO.,
SAN DIEGO, CALIF.

TRUNK IS IN THE FRONT

('47)

('48)

TOWNE SHOPPER
63" WB   (116" LONG)
2 CYL. REAR ENGINE
WT. 600 lbs.   (10.6 H.P.)
ALUMINUM BODY

4.00 x 8" TIRES   50 MPH   40~50 MPG

$595.

Tucker '48

THE TUCKER CORP., 7401 S. CICERO, CHICAGO, ILL. (1946-1949)

PRINCIPAL OUTPUT PRODUCED DURING 1947, BUT KNOWN AS 1948 MODELS, OR EVEN 1949!

REAR.

PRESTON TUCKER
(FOUNDER)
(1903-1956)

SUGGESTED PRICE IN $2500. RANGE (MUCH LESS THAN MFRS. ACTUAL COST!)

6-CYL. 335 CID
HORIZ. OPPOSED
FRANKLIN/TUCKER
REAR ENGINE
(166 HP)

TURNING (CENTER)
"CYCLOPS EYE"
HEADLIGHT

RARE! ONLY 53 BUILT, INCLUDING PILOT MODELS.

SYMBOL OF SAFETY

CRASH COWL

126" WB

Valiant ◭ (1960 ~ 1976)
NEW FROM CHRYSLER

V-100 $2053. (52,788 BLT.)

V100 WAGON (13,946 BLT.) $2365. UP

new INCLINED 6-CYL. 170 CID O.H.V. ENGINE 101 HP @ 4400 RPM or 148 HP @ 5200 RPM

V-200

106½" WB (THROUGH '62) PLYMOUTH'S new COMPACT CAR

V-200

1960

60 QXI-L OR QXI-H new!

V-200s have EXTRA SIDE CHROME TRIM. ➡

$2053. ~ 2566. PRICE RANGE

4-DR. (106,515 BLT.)

$2130.

(25,695 BLT.) 4-DR.

V-100

225 CID PLYMOUTH 6 CYL. ENGINE ALSO AVAIL. (145 HP @ 4000 RPM)

$2423. **V-200** WAGON (10,794 BLT.)

$2014.

new H/T $2137.

V-200

1961

V-200

DASH

V-200

61 RVI-L OR RVI-H

H/T (18,586 BLT.) note GRILLE CHANGE

FINAL YEAR FOR 148-HP VERSION OF SMALL ENGINE

4-DR. (59,056 BLT.)

V-200

$2110.

TOTAL 1961 PRODUCTION: 120,848

CALENDAR YEAR PRODUCTION: 122,275

(SAME ENGINES AS IN 1960)

$2014. ~ 2423. PRICE RANGE

VALIANT

$2285. $2590. (WEST COAST)

V-200 HAS SIDE TRIM

(55,789 BLT.) $2087.

V-100 WAGON

(5932 BLT.)

V-200

$2381.

SVI 62 new GRILLES

(8055 BLT.)

SVI-L (V-100)
SVI-H (V-200)
SVI-P (SIGNET)

(25,586 BLT.)

new SIGNET (has FRONT BUCKET SEATS)
$2230. $2538. (WEST COAST)

VALIANT

SIGNET has ⬇ INSIGNIA on DARK GRILLE

CALENDAR YEAR PROD.: 153,248

$1910. (new 18-GALLON FUEL TANK) (32,761 BLT.)

Valiant V-100 2-door sedan/metallic green

ENGINE

new CONVTS.

106" WB

TRANSMISSION PUSHBUTTONS

R N D 2 1

63 TVI (TOTALLY RESTYLED)

13M·82

SIGNET CVT.

Valiant V-200 4-door station wagon/dark metallic blue

— MODELS :

V-100			V-200			SIGNET	
2 DR. (32,761)	$1910.		2 DR. (10,605)	$2035.		H/T (30,857)	
4 DR. (54,617)	1973.		4 DR. (57,029)	2097.		$2230.	
4 DR. WAGON			CVT. (7122)	2340.		CVT. (9154 BLT.)	
(5932 BLT.)	2268.		4 DR. WAGON (11,147)	2392.		2454.	

CALENDAR YEAR PRODUCTION : 221,677

392

VALIANT

note THAT THIS LATER SERIES CONVERTIBLE has LESS REAR BRIGHTWORK and DIFFERENT DECK EMBLEM FROM "SIGNET" ILLUSTR. ON PRECEDING PAGE

63½

Valiant presents
AMERICA'S LOWEST-PRICED CONVERTIBLE...$2340*

$ 2215. WEST COAST

2-DR. $1921.

V-100
(35,403 BLT.)

Valiant V-100 2-Door Sedan

Barracuda (new) (INTRO. 4-2-64)

$2375. (23,443 BLT.)
6 CYL. OR V8

WEST COAST PRICE $2670. (6)

64

VVI-L (V-100)
VVI-H (V-200)
VVI-P (SIGNET 200)
VVI-P29 (BARRACUDA)

$2388. ↓ V-200

$2256.
$2549. WEST COAST

SIGNET

$2473.

SIGNET CVT. (7636 BLT.)

H/T

new GRILLE with "PLYMOUTH" NAMEPLATE ABOVE

AJ-6830

(37,736 BLT.)

Valiant/64 style
Best all-around compact

$1921. ~ 2473.
PRICE RANGE

ENGINES :
170 CID SLANT-6 (101 HP)
225 CID " " (145 HP)
new 273 CID V8 (180 HP)
@ 4200 RPM

VALIANT

$2487.

BARRACUDA
(64,596 BLT.)

$2801. (V8)
(WEST COAST)

65
new GRILLE

ALL-VINYL
SEATS IN "100."

6 and V8 ENGINE
SPECS. as BEFORE,
EXCEPT THAT new OPTIONAL
10.5 COMPRESSION VERSION of
V8 is ALSO AVAIL., with 235 HP
@ 5200 RPM

$2004.

100

$2476.

$2195.

2-DR.
(40,434 BLT.)

Plymouth Valiant 200 4-Door Station Wagon
(6133 BLT.)

200

$2127.
2-DR.

6 CYL. :
AVI-L (100)
AVI-H (200)
AVI-P (SIGNET)
AVI-P29 (BARRACUDA)
(V8s have "AV2"
PREFIXES)

(41,642 BLT.) 200

(8919 BLT.)

The Roaring
'65s*

$2340.

DASH

Valiant Signet
H/T

new FLAT-PROFILE
AIR CONDITIONER

(10,999 BLT.)

CVT.
(2578
BLT.)

$2561.
$2004. TO $2561. PRICE RANGE

CALENDAR YEAR PRODUCTION :
VALIANT = 139,436
BARRACUDA = 54,855

$2234. TO $2932.
(WEST COAST PRICES)
* = SLOGAN APPLIES TO PLYMOUTH ALSO.

Valiant 66

SIGNET H/T $2261.
$2487. (WEST COAST)

PLYMOUTH DIVISION ◆ CHRYSLER MOTORS CORPORATION

$2556.

6-CYL. OR V8

FROM $2862. (WEST COAST)

↗ Barracuda

100, 200 OR SIGNET MODELS

106" WB

100 = 2025.~2387.
200 = 2226. OR 2502. (SEDAN OR WAGON)
SIGNET = 2261. (H/T) OR 2527. (CVT.)

AVAIL. AS FASTBACK H/T (38,029 BLT.)

TOTAL 1966 PRODUCTION: 138,137 (VALIANT ONLY)

(SINCE 1964)

('67 BARRACUDA INTRO. 11-25-66)

new 108" WB

('67 VALIANT INTRO. 9-29-66)

$2117. 100 2-DR. $2346. WEST COAST

DASH

TAIL-LT.

67

new GRILLE

(29,093 BLT.)

100 (WITH 200 DECOR OPTION) (NOTE SIDE CHROME)

$2163. UP

$2308.

$2262.

↑ 100 REAR

4-DR. $2537.

SIGNET

2-DR. $2491.

'67

TURN SIGNAL INDICATOR IS VISIBLE TO DRIVER

(200 ELIMINATED AS A MODEL SERIES)

TOTAL 1967 PRODUCTION: 108,969 VALIANTS; 62,534 BARRACUDAS (NOW IN 3 BODY TYPES) FASTBACK - CVT. - H/T

Valiant

H/T (19,997 BLT.)

BARRACUDA $2605.

FR. $2936.
(WEST COAST)

**Barracuda.
4 new engines.**

225 CID 6 (145 HP)

318, 340, 383 CID
V8s
(FROM 230 HP)

VALIANT SIGNET
2-DR. $2633.
(WEST COAST)
(6265 BLT.)

6.95 × 14
TIRES (ON
BARRACUDA
SINCE 1967)

(INTRO. 9-14-67)

68

new GRILLES

(FINAL YR. FOR
SPLIT GRILLE ON
VALIANT)

$2400.

6.50/7.00
× 13 TIRES

ENGINES : 170 CID SLANT-6 (115 HP); 225 CID SLANT-6 (145 HP);
273 CID V8 (190 HP); 340 CID V8 (275 HP); 383 CID V8 (330 HP)
(340 and 383 V8s OPTIONAL, BARRACUDAS ONLY)

V-100, SIGNET OR
BARRACUDA

108" WB (SINCE '67)

TOTAL 1969 PRODUCTION:
107,218 VALIANTS ; 31,987 BARRACUDAS

VALIANT

new GRILLES

69

$2313.
SIGNET 4-DR.
$2737.
(23,906 BLT.)

$2813.

'CUDA
340

('CUDA 383 AVAIL.)

(INTRO. 9-19-68)

(17,788 BLT.)

**Barracuda Coupe,
Convertible and 'Cuda**

WILLYS 'Jeep'

WILLYS-OVERLAND, TOLEDO, OHIO

UNIVERSAL JEEP (4-W-D)

MILITARY STYLE ('46)

← $1146.

EARLY MODEL ('46)

CLARK & COMPANY

Station Wagon

All-Steel Station Wagon *(new)* ENGINE →

MILITARY JEEPS INTRO. '41

$1565.

46-47

(JEEP LT. TRUCKS ALSO AVAIL.)

CJ-2A SERIES

('47)

"Jeep Station Wagon" ON HOOD

104" WB

the Jeepster

$1765.

JEEPSTER PRODUCED 1948 TO 1953. REVIVED 1967 BY KAISER JEEP.

(new) →

134.3 CID 4 L-HEAD (63 HP)

ALSO AVAIL. IN 1949: 134.3 CID 4 (F-HEAD, 72 HP) 148.5 CID 6 (72 HP)

48-49

104" WB

(10,326 BLT., 1948)
(1949: 4 CYL. 2307; 6 CYL. 653)
PRICE REDUCED IN 1949.

'Jeep' Station Sedan

6 CYL. only

has IMITATION WICKER PANELING →

(PREVIOUS 4-CYL. STATION WAGON ALSO CONT'D., $1625.)

DEEP UPHOLSTERED SEATS, interior roominess and road-leveling wheel suspension add to the smooth, luxurious riding comfort of the 'Jeep' Station Sedan.

$1865.

6.70 × 15

WILLYS 'Jeep'

JEEPSTER PRODUCTION:
4 CYL. (4066) 6 CYL. (1778)
(1950~1951)

2-W-D WAGON (4 CYL.)
$1783.

($2204. WITH 4-W-D)

('50)
LATER
SER.

('51)

JEEPSTER
$1603. UP $1597. UP
('50)   ('51)

6.50 x 15" TIRES

new
GRILLE (EARLY '50 has '49-STYLE GRILLE.)

50-51

4 CYL. OR 6   OVERDRIVE OPTIONAL

VARIOUS JEEPS, JEEP WAGONS and JEEPSTERS CONTINUE.

MILITARY JEEP

FIRST NON-JEEP WILLYS CAR PRODUCED SINCE 1942 MODELS:

The Revolutionary New Aero Willys

108" WB

AERO-WING
('52 ONLY)

ALL AEROS ARE 2 DR. SEDANS EXCEPT EAGLE (H/T)

AERO-ACE

52

AERO-LARK has 6-CYL. L-HEAD engine; WING, ACE, EAGLE have HURRICANE 6 F-HEAD.   JEEP WAGONS, JEEPSTERS also

DETAILS OF F-HEAD COMBUSTION CHAMBER

New Hurricane 6 Engine. F-head design with 7.6 compression, one of the world's most efficient power plants.

1952 PRODUCTION:	AERO-LARK	(7474)	$1731.	(652-K SERIES, MODEL KA2~675)
31,363 TOTAL	AERO-WING	(12, 819)	1989.	652-L " " LA1~675
(NOT INCL. JEEPS, WAGONS OR JEEPSTERS)	AERO-ACE	(8706)	2074.	652-M " " MA1~685
	AERO-EAGLE	(2364)	2155.	" " " " MC1~685
161 CID 6 (75 HP, LARK; 90 HP, OTHER AEROS.)				CALENDAR YR. PRODUCTION: 35,954

Willys Aero

$1732.

FIFTIETH YEAR
ANNIVERSARY OF
WILLYS-OVERLAND

53

4-DR. AEROS NOW AVAIL.

AERO LARK
4 OR 6 CYL.

(7692 BLT.)

NO HOOD ORNAMENT ON LARK

1646.

2-DR. (8205 BLT.)

AERO-LARK DELUXE →

$1861.

AERO-ACE
$2038.

4-DR. (7475 BLT.)

AERO-FALCON

4-DR. (3117 BLT.)

(REPLACES '52 AERO-WING) (AERO-FALCON 2-DR. RESEMBLES LARK DLX.) (2 DR. ALSO) FALCON and LARK 6 HAVE 75 HP.

$1963.
(4988 BLT.)

AERO-ACE 2-DR.

ACE H/T ALSO AVAIL. (ILLUSTR. AT LOWER RIGHT) (ACE and EAGLE HAVE 90 HP.)

AERO-EAGLE H/T

AERO EAGLE H/T (7018 BLT.)
$2157.

JEEP 4 (4-W-D) ## Station Wagon

JEEP 6 DELUXE
(2-W-D)
SERIES 685

SERIES 4 x 475

$1949.

$2304. ($1862. W/o 4-W-D)
(SERIES 475)

LOW-PRICED AERO-LARK NOW AVAIL. WITH 134.2 CID 4 (72 HP), AS WELL AS WITH THE 161 CID 6 (75 HP) ALL OTHER AEROS w. 6 CYL. ENGINES

TOTAL 1953 PRODUCTION: 42,057

WILLYS, CALENDAR YEAR
35,146

THE NEW 1954 *Aero* WILLYS

(LARK AVAIL. WITH 134 CID 4 OR 161 OR 226.2 CID 6)

K-W 1954 KAISER/ WILLYS MERGER
Kaiser-Willys Sales Div. Willys Motors, Inc.

(11 BLT.)
REAR SPARE TIRE
AERO EAGLE CUSTOM

27% MORE POWER (115 90 OR 72 HP)

LARK, ACE OR EAGLE MODELS
$2222.

AERO FALCON NO LONGER AVAIL.

54

NEW HOODED HEADLTS., NEW SEGMENTED TAIL LIGHTS NEW HUBCAPS
$1737. UP

EAGLE DE LUXE H/T (84 BLT.)

4 OR 6 CYL.

WAGON has new GRILLE

PRODUCTION = 11,856 CALENDAR YEAR 9339

WIDER OPENING TOP DOOR GIVES WIDEST OPENING OF ANY STATION WAGON IN ITS FIELD

4x415 $2304.

6 CYL. 2WD 4WD
$1973./2399.

SER. 685

1955 WILLYS PRODUCTION = 6565 CALENDAR YEAR = 4778
1955 AERO MODELS CUSTOM (3112 BLT.); ACE (659 BLT.); BERMUDA (2215 BLT.)
2 OR 4 DR. 4-DR H/T

ENGINES: 161 CID 6 (90 HP) OR 226.2 CID 6 (115 HP) IN ACE; OPT. IN CUSTOM OR BERMUDA

MANY JEEPS WITH FOUR-WHEEL-DRIVE CAN BE RECOGNIZED BY PLAIN WHEEL HUBS (BELOW)

$1795. 4-DR.
(2822 BLT.) CUSTOM

4 OR 6 CYL.

NO HUBCAPS ON THIS 4-W-D WAGON

4-W-D

all-new GRILLE, TAIL- LIGHTS and TRIM

55

"AERO"

FINAL WILLYS CAR BUILT IN U.S.A., BUT JEEP PRODUCTION CONTINUES. KAISER JEEPS BLT. 1963 TO 1969. AMERICAN MOTORS CORP. BEGAN BUILDING JEEPS (SINCE START OF 1970 MODEL SEASON.) MFD. BY CHRYSLER CORP. SINCE 1987.

BERMUDA H/T
$1997.